DOMESTIC 007:

ELIMINATING HIDDEN KILLERS IN YOUR HOME

RIDDING YOUR HOME OF HARMFUL CHEMICALS
AND CANCER-CAUSING AGENTS

By

Maryla Webb Records

Published in the USA
by
Earth-Wise Publications
Silver Spring, Maryland

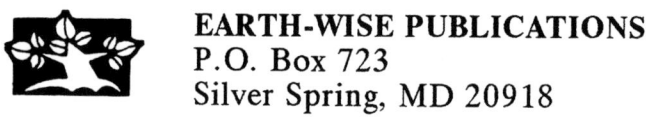

EARTH-WISE PUBLICATIONS
P.O. Box 723
Silver Spring, MD 20918

Records, Maryla Webb, 1953 -
   Domestic 007: eliminating hidden killers in your home : ridding your home of harmful chemicals and cancer-causing agents/ by Maryla Webb Records.
         p.   cm.
   Includes bibliographical references.
   ISBN 0-9655793-0-1 (pbk.)
   1. Housing and health. 2. Household supplies–Toxicology.
I. Title.
RA770.5.R43 1996   96-38137
615.9'02--dc21   CIP

Cover design by Jim and Tia Gray. Photography by John De Fabbio Photographer.

Attention environmental groups, health organizations, educational groups, universities and others: This book is available at special quantity discounts for bulk purchases for educational purposes, sales promotions, fund-raising or gifts. For information, write the Special Markets Department, Earth-Wise Publications, P.O. Box 723, Silver Spring, MD 20918. Phone: (301) 681-3492, FAX: (301) 681-9892. Also see the order form at the back of the book.

Disclaimer: Please Read This!

This book is dedicated to the earth and all its inhabitants. May we live in health, peace and harmony, and prosper by living in accord with God's and nature's laws.

It is also dedicated to those who have died of cancer, including many friends and family: Grandfather Thomas William Morgan (throat), Aunts Carolyn Morgan Wallace (breast), Mary Ruth Morgan Coleman (pancreas), Virginia Morgan Haskell (breast), Grandfather John Walter Webb Sr. (stomach), Aunt Katherine Webb Urquhart (pancreas), father-in-law Lee Maloney Records (lung and liver), Aunt's mother, Vera Brent Smith (pancreas); long-term family friends John Wade (lung), Katherine Praytor (breast), Edwin Webb (stomach), Carolyn Nolan (breast), Jack Till (prostate), Maurice Kahn (lymphoma), and Phillip Robinson (leukemia) among others; and adult friend, Mary Horne (breast). If this book helps even one person avoid cancer, it would not have been written in vain. May God bless us all and lead us to unravel any remaining mysteries or obstacles to the elimination of cancer and the restoration of health.

Acknowledgements

I thank my parents, John and Cecile Webb, who instilled in me the knowledge that I could do anything I wanted and the persistence to finish any task. I also thank my husband for his loving support, particularly his financial support for the couple years that it took to complete this book. I thank Karen Dix for introducing me to the topic of chemical sensitivity; Bonnie Fisher and Patricia Gokey, my wonderful ex-housemates, who taught me much about living without toxins; and Ed Passerini, the University of Alabama professor of environmental studies who first ignited my interest in environment. I would like to thank Carol Merrill of "Let's Make a Deal" for her encouragement and excitement about the book; Scott Pollick (PHD Dog Food) and Dr. Monique Maniet (Holistic Vet, Bethesda, Maryland) for their helpful comments on the pet sections; Atty. Raymond Konan for his helpful information on meat and health; Mothers for Natural Law for their information on genetically engineered food; Eric Rivkin (New Leaf, Minnesota) for his help on the complicated issue of plastics; by Dr. Michael Atkinson (Naturopath, Maryland) for his helpful comments on the cooking section; and the review by Ms. Susan MacKenzie (Healthy Spaces, Virginia). A number of people at the EPA were particularly helpful: Jerry Blondell of the Health Effects Division, Office of Pesticide Programs; Debbie Janes, Public Affairs Office at the Environmental Research Center, Research Triangle Park; and Janet Remmers, Office of Pollution Prevention and Toxics, Chemical Management Division. I also thank the authors whose ground-breaking research and efforts greatly facilitated the writing of this work, particularly, Deborah Lynn Dadd, Ellen Greenfield, Aubrey Hampton, Linda Mason Hunter, and David Steinman and Samuel Epstein. (See *References*.)

The final cover design was created by computer artist Jim Gray at jwgray@erols.com. (See his WEBsite and incredible computer art at http:// www.erols.com/jwgray/.) Tia and Jim Gray of Alexandria, Virginia, provided the illustrations. The author's photographer was John De Fabbio in Silver Spring, Maryland at defabbio@erols.com. (See his WEBsite of beautiful fine art at http://www.johndefabbiophoto.com.) The talented Thai Tran of Beauty by Thai, Arlington, Virginia, created the hair styling and make-up. The editor was the notable Cynthia Lane of Des Moines, Iowa.

Finally, I extend my deepest gratitude to Maharishi Mahesh Yogi, who only knew he had something the world needed, and made available the powerful Transcendental Meditation (TM) technique for the development of human potential and higher states of consciousness.

Table of Contents

Preface   vi
Chief Seathl's Letter of 1855   viii
Checklist Summary   x

I. House Structure

    A. Insulation   1
    B. Wood Products   3
    C. Radon   3
    D. Water Pipes   3
    E. Water Spouts or Valves   3
    F. Fireplaces/Wood Stoves   4
    G. Roof   4
    H. Garage   4
    I. Home Design   4

II. Home Interiors

    A. Flooring   5
    B. Walls   7
    C. Ceilings   8
    D. Lighting   8
    E. Electricity   9
    F. Water   10

III. Home Furnishings

    A. Commercial Furniture   11
    B. Mattresses   11
    C. Drapes and Bedclothes   13
    D. Toys   13

IV. Appliances

    A. Gas Appliances   15
    B. Microwaves   15
    C. Vacuums and Dusting   16
    F. Humidifiers   16
    D. Clothes Dryers   16
    E. Kerosene Heaters   16

V. Consumables

    A. Aerosols   17
    B. Cleaners   17
    C. Personal Care Products   19
    D. Paper Products   20

V. Consumables (continued)

    E. Air Fresheners   20
    F. Tobacco   20
    G. Car Care   21
    H. Pest Control   22
    I. Lawn/Garden Care   24
    J. Pet Care   25

VI. Food

    A. Synthetic Chemicals   27
    B. Pesticides   27
    C. Irradiation   29
    D. Cookware and Dishes   29
    E. Packaging and Storage   29
    F. Beverages   30
    G. Salt   31
    H. Oils   31
    I. Meat   32
    J. Genetic Engineering   33
    K. Sweeteners   34
    L. Water   35
    M. Pet Food   36

Appendices   37

Product Sources   38
Further Resources   44
Muscle Testing/Edu-K   49
Recipes   50
Energy & Nutritional Supplements   52
Liver & Gallbladder Detoxification   53
Cancer and Modern Living   54
Aging and Modern Living   56
Air and Air Purification   58
Negative Ions and Health   60
Water and Water Purification   61
Electromagnetic Fields (EMFs)   67
Health Effects of Toxic Substances   68
Cosmetic Ingredients to Avoid   71
Disposal of Hazardous Wastes   72
Hazardous Waste Disposal Chart   73
Labeling and Regulation   74
Natural Lawn and Garden Care   75
General Environmental Issues   76
Cosmic Consciousness   78
Inner Housecleaning   80
References   81

Preface

When I look back, my initial interest in home health seems to have unfolded naturally and in stages over a number of years. My first interest in environment sprang from my Alabama upbringing close to a lake and the woods. In college, the environmental movement gave me an awareness of the issues and, with existing feelings of connectedness, ignited a desire to work for the environment. This culminated in graduate work in ecology at Yale and an environmental career. My primary work and interest lie in international conservation of natural resources. However, I care about all the areas of environment and their interconnections, particularly having been also a long-term student of Eastern philosophy and meditation. I see nature holisticly – that actions are either in tune with nature and her laws, or attempts to "subdue" nature and contravene those laws. The latter actions are doomed to failure in the long run, as nature gives us warning signals or takes her own corrective actions. The spread of cancer and chemical sensitivities may be two of these warnings or corrective actions.

Cancer has struck in my immediate family and circle of friends many times, often fatally. I wondered about the general rates. When researched it, I discovered that cancer now strikes one in three persons and kills one in four, up forty percent since 1950, the dawn of the petrochemical age, processed foods, and the spread of the peaceful use of the atom. (See *Cancer and Modern Living* in the Appendix.)

I meet more and more people with allergies and chemical sensitivities. Karen Dix of Karen's Nontoxics is one of those people. Karen experienced a severe illness due to a chemical overload resulting from a simple home project. Years of repeated visits to doctors led her nowhere. Finally, she diagnosed herself and took back her life by detoxification and careful control of the day-to-day substances she contacts. Ms. Dix now helps others with similar problems and finds that simple changes can resolve even serious illness like lymphoma or chronic arthritis.

I cannot help but think that the spread of chemical sensitivities and cancer are connected to the now ubiquitous use of synthetic chemicals. In 1940, the U.S. manufactured one billion pounds of synthetic chemicals per year. By 1950, that figure rose to fifty billion; and by 1980, it had reached 500 billion pounds. Many of these chemicals are used in household products and end up in our homes and our bodies. For example, when a Harvard scientist did a five-year study for the EPA, he found levels of toxic pollutants in homes were often a hundred times higher than those outdoors. The location of the homes didn't change the data. Homes in pristine rural areas had the same levels of pollution as homes next to chemical factories or smelter plants. EPA researchers also find residues of more than 400 toxic chemicals in human blood and fat tissue. (See *Aging and Modern Living* in the Appendix.)

As Pogo states so well in his comic strip, "We have seen the enemy and he is us." We put our health at risk in our own homes through the materials we use to construct them and furnish them, as well as those we use to keep our houses, ourselves and our possessions clean and pest- and odor-free. We will send the strongest message to manufacturers about our preference for life-supporting products by altering our buying patterns. In many cases, it is not more expensive to do so. Only by buying safe products using natural materials can we cease to pollute our air, water, homes and bodies. Only through our own actions as consumers can we change the world, one household at a time.

To educate myself on how I could protect my family's health, I began to read all the books I could find on these topics. Each book was helpful in its particular field, but no single book covered all the topics of interest to me. Some of the books were already outdated and none covered the topics in the succinct, overview fashion I needed in these busy times. I imagined that I could easily produce such a book and naively began to write. A couple years later, you have the results.

Amazingly, all my research condensed into a book whose actual text is only thirty pages long, but which manages to cover everything from insulation to pet food. This compact guide not only gives an overview of potential dangers but also gives solutions, corrective actions and alternatives. When you use this book, please remember that I do not pretend to be an expert in any of the many areas which it covers. Anyone could devote a lifetime of study to each of these topics, and the field continues to change rapidly. I tried to find the best information I could in a reasonable time, but I did not conduct independent studies or an exhaustive review of the primary literature (scientific publications). Therefore, I cannot guarantee the accuracy of absolutely everything. The publisher and I therefore make no representations about and cannot be responsible for incorrect or incomplete information or any adverse health or environmental effects caused by any product or process. Think of this guide as a starting point to conduct your own research and come to your own conclusions in order to make responsible choices.

Although this book contains many practical tips for creating a healthier home environment, I believe that our environmental problems also require inner work. It's possible – as claimed by many saints, sages and indigenous peoples both now and throughout the ages – that our environmental crisis is at root a spiritual one. (See *General Environmental Issues*, Appendix.) Somehow, in the industrial and scientific age and the age of rational economics, we have lost our sense of awe, reverence and connectedness to each other, to nature and to God. (See *Chief Seathl's Letter of 1855*.) Environmental impact analysis, environmental economics, governmental laws and incentives, scientific analysis and technological fixes are insufficient. Ultimate success also depends on reawakening the spiritual awareness and values which recognize the wholeness of life and uphold nature's laws.

One meditation technique that I find useful in the growth of awareness is Transcendental Meditation (TM). Although many helpful forms of meditation exist, the TM technique is the most researched form on the planet – with 500+ published studies – due to its consistent results from subject to subject. TM rapidly brings about higher states of consciousness, greater awareness of our connectedness and a deepening of mental, emotional and spiritual capacities.

I hope you find the book useful and interesting. If you have any questions or comments, please write me c/o Earth-Wise Publications, P.O. Box 723, Silver Spring, MD 20918 or Email: 75734.1202@compuserve.com.

Chief Seathl's Letter of 1855

The Great Chief in Washington sends word that he wishes to buy our land. The Great Chief also sends us words of friendship and good will. This is kind of him, since we know he has little need of our friendship in return. But we will consider your offer, for we know if we do not so, the whiteman may come with guns and take our land. What Chief Seathl says, the Great Chief in Washington can count on as truly as our white brothers can count on the return of the seasons. My words are like stars – they do not set.

How can you buy or sell the sky – the warmth of the land? The idea is strange to us. Yet we do not own the freshness of the air or the sparkle of the water. How can you buy them from us? We will decide in our time. Every part of this earth is sacred to my people. Every shining pine needle, every sandy shore, every mist in the dark woods, every clearing and humming insect is holy in the memory and experience of my people.

We know that the whiteman does not understand our ways. One portion of the land is the same to him as the next, for he is a stranger who comes in the night and takes from the land whatever he needs. The earth is not his brother, but his enemy, and when he has conquered it, he moves on. He leaves his fathers' graves behind and he does not care. He kidnaps the earth from the children. He does not care. His fathers' graves and his childrens' birthright are forgotten. His appetite will devour the earth and leave behind only a desert. The sight of your cities pains the eyes of the redman. But perhaps it is because the redman is a savage and does not understand.

There is no quiet place in the whiteman's cities. No place to hear the leaves of spring or the rustle of insect's wings. But perhaps because I am a savage and do not understand – the clatter only seems to insult the ears. And what is there to life if a man cannot hear the lovely cry of a whippoorwill or the arguments of the frogs around a pond at night? The Indian prefers the soft sound of the wind darting over the face of the pond, and the smell of the wind itself cleansed by a mid-day rain, or scented with a pinon pine. The air is precious to the redman. For all things share the same breath – the beasts, the trees, the man. The whiteman does not seem to notice the air he breathes. Like a man dying for many days, he is numb to the stench.

If I decide to accept, I will make one condition. The whiteman must treat the beasts of this land as his brothers. I am a savage and I do not understand any other way. I have seen a thousand rotting buffalos on the prairies, left by the whiteman who shot them from a passing train. I am a savage and I do not understand how the smoking iron horse can be more important than the buffalo that we kill only to stay alive. What is man without the beasts? If all the beasts were gone, men would die from great loneliness of spirit, for whatever happens to the beasts also happens to man. All things are connected. Whatever befalls the earth befalls the sons of the earth.

Our children have seen their fathers humbled in defeat. Our warriors have felt shame. And after defeat, they turn their days in idleness and contaminate their bodies with sweet food and strong drink. It matters little where we pass the rest of our days — they are not many. A few more hours, a few more winters, and none of the children of the great tribes that once lived on this earth, or that roamed in small bands in woods, will be left to mourn the graves of a people once as powerful and hopeful as yours.

One thing we know which the whiteman may one day discover. Our God is the same God. You may think now that you own him as you wish to own our land. But you cannot. He is the God of man. And his compassion is equal for the redman and the white. The earth is precious to him. And to harm the earth is to heap contempt on its creator. The whites, too, shall pass — perhaps sooner than other tribes. Continue to contaminate your bed, and you will one night suffocate in your own waste. When the buffalo are all slaughtered, the wild horses all tamed, the secret corners of the forest heavy with the scent of many men, and the view of the ripe hills blotted by talking wives, where is the thicket? Gone. Where is the eagle? Gone. And what is it to say good-bye to the swift and the hunt, the end of living and the beginning of survival.

We might understand if we knew what it was that the white man dreams, what hopes he describes to his children on long winter nights, what visions he burns into their minds, so that they will wish for tomorrow. But we are savages. The white man's dreams are hidden from us. And because they are hidden, we will go our own way. If we agree, it will be to secure your reservation you have promised. There perhaps we may live out our brief days as we wish. When the last redman has vanished from the earth, and the memory is only the shadow of a cloud moving across the prairie, these shores and forest will still hold the spirits of my people, for they love this earth as the newborn loves its mother's heartbeat. If we sell you our land, love it as we have loved it. Care for it, as we've cared for it. Hold in your mind the memory of the land, as it is when you take it, and with all your strength, with all your might, and with all your heart — preserve it for your children, and love it as God loves us all. One thing we know — our God is the same God. This earth is precious to him. Even the white man cannot be exempt from the common destiny.

Source: Wes Jackson, 1979. *Man and the Environment*. Dubuque, Iowa: William C. Brown Company Publishers, pp. 134-135.

Checklist Summary

House Structure and Design

[ ] Was asbestos insulation used in your home?

[ ] Is your home insulated with fiberglass, mineral wool or ceramic-based fibers?

[ ] Was urea-formaldehyde foam insulation (UFFI) used in your home?

[ ] Was plywood or particleboard used extensively in construction of walls, subflooring, countertops (covered with plastic laminate) and kitchen or bath cabinents?

[ ] Is your house sited in an area with high naturally occurring levels of radon 222 in rocks and soil, or were radon rich materials used in the concrete, tile or brick of your home?

[ ] Is the water main to your house made of lead or do your pipes have lead solder?

[ ] Are your water pipes made of plastic polyvinyl chloride (PVC) or old, galvanized iron?

[ ] Do you have pressure-balanced mixing valves or a child-safety tub spout?

[ ] Does your home contain a fireplace or wood-burning stove?

[ ] Is asbestos tile used on the roof or as siding?

[ ] Does your home have an attached garage?

[ ] Was your home designed using ecological or environmental principles?

Home Interiors

[ ] Is the major part of your floors carpeted or covered with rugs?

[ ] Is your floor covered with synthetic parquet, sheet vinyl or synthetic rubber?

[ ] Are your walls covered with wallpaper, particularly vinyl or self-adhesive wallpaper?

[ ] Are your walls painted with oil-based paints?

[ ] Are your walls painted with lead-based paints?

[ ] Are your walls paneled?

[ ] Are any of your ceilings covered with asbestos tile?

[ ] Are any of your ceilings covered with acoustic tile?

[ ] Do you use fluorescent or incandescent bulbs to light your home?

[ ] Do you have small children and unprotected electric wall outlets?

[ ] Is your house located near high-voltage electric lines, radio or radar towers, microwave relay towers or an electricity generating plant?

[ ] Do you use unpurified water in your home?

Home Furnishings

[ ] Is your home largely furnished with new, commercially produced furniture, particularly pieces made from particleboard (in place of wood), synthetic stuffings and polyester fabrics?

[ ] Is your mattress or pillow made with polyurethane and treated with a flame retardant?

[ ] Do your sheets, blankets, bedspreads and drapes contain polyester or acrylic, or are they permanent press?

[ ] Do your children have many toys made from vinyl plastic containing polyvinylpyrrolidone (PVP) or polyvinyl chloride (PVC), or art or hobby supplies containing toxic materials?

Appliances

[ ] Do you have a gas furnace, stove or hot water heater?

[ ] Do you heat or cook with a microwave?

[ ] Do you have a portable vacuum?

[ ] Do you use a cool-mist humidifier, especially in the nursery?

[ ] Do you have a gas dryer, an asbestos dryer belt or a poorly vented dryer?

[ ] Do you use a kerosene heater in winter?

Consumables

[ ] Do you use commercial products dispensed by an aerosol spray?

[ ] Do you have a house full of commercial cleaning products?

[ ] Do your personal care products contain synthetic chemicals?

[ ] Do you consume paper products bleached with chlorine or containing artificial fragrances or dyes?

[ ] Do you use an aerosol or other artificial or perfumed air fresheners?

[ ] Does someone in your home smoke?

[ ] Do you use commercial products for cleaning and protecting the interior and exterior of your car?

[ ] Does your anti-freeze contain ethylene glycol?

[ ] Do you know how to properly dispose of automotive products?

[ ] Do you use commercial products to control insect pests in the home?

[ ] Do you use a commercial insect repellent containing DEET?

[ ] Do you have a termite bond with a professional termite exterminator?

[ ] Do you use artificial fertilizers or pesticides, or a lawn care company that does?

[ ] Do you use commercial cat litter with silica or artificial perfumes?

[ ] Is your pet bed made with polyester or stuffed with styrofoam?

[ ] Do you use standard, chemical flea or tick collars, dips, or sprays?

Food

[ ] Do you eat a large proportion of processed foods and meat containing food additives and preservatives?

[ ] Do you know which fruits and vegetables have fewer pesticides or do you buy organic produce?

[ ] Do you eat foods that have been irradiated?

[ ] Do you use aluminum or stainless steel cookware or pottery that was imported or made before 1970?

[ ] Do you purchase or store foods in styrofoam, cling wrap, plastic wrap or plastic (PET) containers?

[ ] Do you consume a large amount of coffee or decaffeinated coffee?

[ ] Do you drink bottled water?

[ ] Do you drink more than one alcoholic beverage a night?

[ ] Do you drink many sodas?

[ ] Do you consume large quantities of table salt or processed foods high in sodium?

[ ] Do you use soy or canola oil?

[ ] Does a large proportion of your diet consist of red meat?

[ ] Do you eat genetically engineered food?

[ ] Do you eat much NutraSweet (aspartame) or sugar (sucrose)?

[ ] Do you drink or cook with unpurified tap water?

[ ] Do you feed your pet commercial dry or canned pet food?

Chapter 1   HOUSE STRUCTURE

We spend most of our time in building structures, and they can be a great health hazard when built or filled with many materials available today. Proper siting and design are important factors in safeguarding us from radon and carbon monoxide, and ensuring adequate ventilation while maintaining sufficient insulation. Even if your house was not built of natural materials and designed to protect the health of its inhabitants, you can take steps to reduce dangers in most any situation.

⧴ INSULATION

[ ] *Was asbestos insulation used in your home?* Asbestos insulation may take the form of (i) whitish material wrapped around hot water and steam pipes; (ii) chalk-colored flat or corrugated paper around ducts and furnaces; and (iii) two inch thick boiler wraps. Asbestos releases microscopic fibers that can become permanently embedded in the lung, chest lining, or stomach, and cause cancer. Asbestos is dangerous only if it is damaged so that fibers are getting into the air.
Solution: Insulate with natural cork (R value 4/in.) or Air Krete, a nontoxic, magnesium-oxide-based foam (R value 3.9/in). Have existing asbestos identifed by an experienced contractor or professional laboratory. Removal should always be handled by a professional asbestos removal contractor. A dust mask is useless.
Corrective Actions: Sometimes, leaving the asbestos in place and sealing provides the best health protection at the most reasonable cost. For example, repair frayed pipe insulation by wrapping with good, wide duct tape and sealing with Safe Seal. First spray with soapy water. Wear protective clothing and an asbestos-approved respirator. In addition, use negative ion generator which will cause respirable-size particles to drop to the ground.

[ ] *Is your home insulated with fiberglass, mineral wool or ceramic-based fibers?* Some scientists are concerned that any fiber-based insulation with small, inhalable fibers may pose dangers similar to asbestos. Size and dimension, not substance, are apparently the most important factors in whether a fiber can be inhaled and lodge in lung tissue. Most fiberglass fibers dissolve in animal tissue (unlike asbestos) and are too large to be inhaled, but technology is now making fibers smaller and more durable.
Corrective actions: Seal wherever possible (see asbestos) and block access to attic and basement crawl spaces, particularly to small children. Use negative-ion generation and air filtration. (See *Air Purification* in Appendix.)

[ ] *Was urea-formaldehyde foam insulation (UFFI) used in your home?* The foam can be easily blown into the walls of buildings where it hardens within minutes and cures within days. Formaldehyde leaches from the product into homes, often causing acute illness. Thousands of complaints from residents in homes fitted with this insulation caused the Consumer Product Safety Commission to ban the product in 1982. The ban was overturned in 1983, but the controversy remains.
Corrective action: Air purification. (See *Product Sources* in Appendix.)

1

2

⊢ WOOD PRODUCTS

[ ] *Was plywood or particleboard used extensively in construction of walls, subflooring, countertops (covered with plastic laminate) and kitchen or bath cabinents?* Plywood is made of thin sheets of wood glued together with urea-formaldehyde resin. Particleboard is made by saturating wood shavings with UF resin and pressing the resulting mixture into the desired form. These products outgas large amounts of formaldehyde during the first months or even years of use. In weathertight housing, levels of formaldehyde may exceed safe limits. *Solution:* Use natural wood products. *Corrective actions:* Seal particleboard or plywood panels, pieces or cabinets with Safe Seal. Air purification.

⊢ RADON

[ ] *Is your house sited in an area with high naturally occurring levels of radon 222 in rocks and soil, or were radon-rich materials used in the concrete, tile or brick of your home?* Radon can migrate from dirt and rock and enter your home. Radon-decay products emit high levels of alpha radiation, attach themselves to respirable particles and may lodge deeply in lung tissue, causing cancer. *Solution:* Don't buy or build in areas with naturally high radon levels. Particularly avoid areas of uranium or phosphate mining where tailings have been used for landfill or in concrete. *Corrective actions:* Inexpensive radon detectors are available in stores. If high levels are found, seal all cracks in basement walls and concrete slabs with caulk, then paint with a sealant like Safe Seal. Ventilate crawl spaces to dispel gas. Air purification or filtration reduces radon progeny. In serious cases, call a professional.

⊢ WATER PIPES

[ ] *Is the water main to your house made of lead or do your pipes have lead solder?* Lead pipes and lead solder are sources of lead in water.
[ ] *Are your water pipes made of plastic polyvinyl chloride (PVC) or old, galvanized iron?* PVC plumbing allows polyvinyl chloride and other carcinogenic compounds to leach into water. Moreover, joints sealed with solvent-based glue are linked to birth defects in lab animals and lymphoma. Old, galvanized pipe can leach cadmium, which replaces the nutrient zinc in the body. Cadmium is suspected of inhibiting the immune system and causing other illnesses. *Solution:* The best piping is copper, although even excess amounts of copper is toxic. Use solder that is both lead- and antimony-free. One noted author states that PVC piping is contaminant-free after three weeks of flushing. Don't drink water from the hot water tap, as contaminants collect in the tank. Testing is recommended. *Corrective actions:* Water purification.

⊢ WATER SPOUTS OR VALVES

[ ] *Do you have pressure-balanced mixing valves or a child-safety tub spout?* Each year more than 37,000 children under the age of fourteen are treated for burns caused by scalding hot water, often in the bathtub. *Solution:* Turn your hot water heater down below 120° (where scalding occurs) to a comfortable temperature. *Corrective Actions:* Make sure you have a good pair of pressure-balanced mixing valves or temperature-sensitive tub spout. Cheaper than mixing valves, the tub spout actuator shuts off water flow when the temperature reaches 114° F. A red restart button resets the actuator once the water is tempered.

✦ FIREPLACES/WOOD STOVES

[ ] *Does your home contain a fireplace or a wood-burning stove?* Wood smoke produces a large variety of air pollutants, including carbon monoxide, nitrogen oxides and organic chemicals, including carcinogens, most within the respirable size range. *Corrective actions:* Ensure that exhaust pathways are clean and do not leak. Each device should have a separate flue spaced apart. Do not burn colored print paper or synthetic fire logs. Close the fireplace opening with glass doors. Buy a wood stove with a secondary combustion chamber or a smokeless stove. Air purification.

✦ ROOF

[ ] *Is asbestos tile used on the roof or as siding?* This material was used after World War II as facade material. It can be recognized by wavey, yet parallel scoring used over most of the surface to impart texture. Though generally hard and brittle, it looks fibrous around the edges if broken. As the tiles weather, they begin to shed asbestos fibers which may enter the home through doors and windows. *Solution:* Have the tiles identified and replaced by a professional. *Corrective Action:* Air filtration.

✦ GARAGE

[ ] *Does your home have an attached garage?* Running a car in an attached garage pollutes a home with car exhausts. Carbon monoxide, nitrogen dioxide, volatile organic chemicals and lead are some of the pollutants. *Solution:* The best solution is to have an unattached garage. If you prefer an attached one, isolate it well away from living areas and air handling ducts. Also store hazardous materials away from the house in a locked shed. *Corrective actions:* Do not start the car with the garage door closed and close off any ducts leading from the garage to the house. Insulate well any adjacent walls and ceilings. Use air purification and/or negative ion generation in adjacent living areas.

✦ HOME DESIGN

[ ] *Was your home designed using ecological or environmental principles?* For those designing and building a new home, many resources are available to aid you in ensuring an *ecological* design. *Baubiologie* from West Germany emphasizes the use of natural materials and the impact of buildings on the living environment. *Feng shui* and the Western equivalent, *Geomancy*, are concerned with the harmonious placement of buildings and arrangement of rooms and furnishings. The Vedic system of sacred siting and building placement, *vastuvidya*, has been revived by Maharishi Mahesh Yogi as *sthapatyaveda*. Even if you are just buying a house, you may want to consider whether the house has serious problems from one or more of these perspectives. Techniques are available for correcting some existing problems. For further information on one of these disciplines, contact:
(1) International Institute for Baubiologie and Ecology, P.O. Box 387, Clearwater, FL 34617 (813) 461-4371
(2) The National Association of Realtors frequently knows who does *feng shui* consultations as it may affect a house sale with Asians.
(3) Maharishi Sthapatya-Veda P.O. Box 272
NL 6300 AG Valkenburg
The Netherlands
FAX #011-31-04406-13262

4

Chapter 2 HOME INTERIORS

Many of the things we use to panel, paint, floor, carpet and furnish our homes contain synthetic and toxic chemicals that outgas, causing potentially unsafe levels of air pollution, particularly in airtight homes. Fortunately, in most cases, we can take helpful actions short of throwing everything out and starting again.

✈ FLOORING

[ ] *Is the major part of your floors carpeted or covered with rugs?*
I make no distinction between synthetic and natural fiber rugs because any kind of carpet, particularly old carpet, is a haven for microorganisms (molds, fungi, bacteria, dustmites). Studies have measured 10 million micro-organisms per square foot of carpet. In addition, toxic chemicals outgas from synthetic carpet, carpet pads and adhesives. Even natural fiber carpets and rugs may outgas toxic chemicals if colored with toxic dyes, if backed with synthetic latex (which outgases 4-PC), or if grown or subsequently treated with pesticides. New carpet may release as many as thirty chemicals into the air, although most outgassing occurs in the first three months. (Formaldehyde has now been banned for use in carpet manufacture.)
Solution: Use hardwood, brick, marble, terrazzo, ceramic or stone tile, cork, natural or battleship linoleum flooring. When finishing these floors, use nontoxic adhesives and paints, and stains and waxes from plant ingredients. (See *Product Sources.*) If you carpet, use natural, untreated fiber rugs or wall-to-wall carpeting. If wool wall-to-wall is too pricey, synthetic nylon is the least risky synthetic. Try to get untreated, natural latex or jute backing.

Corrective actions: Although cleaning a carpet will not have any effect on the outgassing of new carpet, regular cleaning (every six months) can reduce micro-organisms and make carpet as healthful as a hardwood floor. Use a IICRC certified firm using a hot-water extraction method. (See *Further Resources.*) Spray or wash a new carpet with a mild vinegar solution (Solution #1, see *Recipes*) and allow it to air before installing. Already installed synthetic carpets should be sprayed after cleaning every six months. If vapors are still a problem, you may want to seal the carpet with AFM Carpet Guard. Air purification and negative ion generation reduces toxic chemicals and micro-organisms. Vacuuming pulls micro-organisms to the upper layers of carpet; and air exhausted from the vacuum's dust bag suspends the smaller dust particles and micro-organisms. (See *Appliances.*)
[ ] *Is your floor covered with synthetic parquet, sheet vinyl, or synthetic rubber?* Wood parquet floors are likely to contain harmful adhesives. Likewise, sheet vinyl and rubber floors probably contain toxic chemicals, waxes or adhesives that outgas and produce odors.
Solution: Buy prefinished, hardwood floor tiles (six inch squares held together with wire) and lay them with nontoxic white or yellow glue. A few manufacturers offer natural linoleums.
Corrective action: Air purification.

5

⟶⊢ WALLS

[ ] *Are your walls covered with wallpaper, particularly vinyl or self-adhesive wallpaper?* Many wallpapers are treated with fungicides and pesticides which outgas after installation. The materials composing vinyl papers and wallpaper adhesives also contain many toxic chemicals that outgas. Vinyl and self-stick wallpapers are considered the worst offenders.
Solution: Paint rather than wallpaper walls. Use a water-based latex paint; buy natural and nontoxic paints from specialized companies; or mix your own. If you must wallpaper, request a materials safety data sheet from the wallpaper manufacturer. Use nontoxic adhesives.
Corrective actions: Flaking paper or wallpaper with detectable odors should probably be removed and replaced with paint (particularly in an infant's room). Wash other paper with a baking soda solution. (See *Recipes.*) Air purification.

[ ] *Are your walls painted with oil-based paints?* A John Hopkins University study found over 300 toxic chemicals and 150 potential carcinogens may be present in paint. Oil-based paints are the worst as they contain petroleum-based organic solvents that are volatile but outgas very slowly. Persons in rooms painted with oil-based paints will inhale these toxic vapors for months. Toxic vapors from exterior walls can affect plants. Even water-based latex paints may contain pesticides and fungicides, but will cure much quicker.
Solution: Paint with natural and nontoxic paints from specialized companies or mix your own. (See *Product Sources* and *Recipes.*) To neutralize pesticides in water-based latex paints, add baking soda to the paint until it doesn't bubble anymore or add 2 oz. pure vanilla extract.
Ccrrective actions: Wash down painted walls with Vinegar Solution #1 and baking soda. (See *Recipes.*) Air purification.

[ ] *Are your walls painted with lead-based paint?* Older homes built before 1950 are most likely to contain lead-based paint. Before then, lead compounds were added to interior and exterior house paint to make it shinier and more durable. Though levels were lowered in the 1950s, lead was added until as late as 1976. As this paint peels and chips, it can become a health hazard. Tiny breathable flakes can become airborne, and lead can leach out of paint and wash into the soil around your home. Once soil is contaminated, lead remains for 2,000 years, mostly in the top one half inch.
Corrective actions: If in doubt, have a paint sample tested. (See *Further Resources, Testing.*) If lead is found, great precaution should be taken in its removal. If no paint is loose, coverage with a latex paint (see above) and Safe Seal may be preferable to removal. Removal is a dangerous task, best left to highly reputable professionals. Ask the professionals how they plan to conduct the removal. The best method is a combination of chemicals and scraping with a sharp blade. Heat guns, blowtorches, power sanders and sandblasters either vaporize the lead or create fine breathable dust. Clear anyone, including pets, from the job site; remove all belongings; seal all permanent fixtures with plastic sheeting and tape; and use a window or other forced air exhaust system. Outside jobs require drop cloths. Close windows and remove all toys from the area. Vacuum afterwards with an industrial vacuum. Contact the board of health for proper disposal.

[ ] *Are your walls paneled?* Most commercial paneling today usually consists of three plies of wood veneer glued together with urea-formaldehyde resin. This product has a high formaldehyde content and outgassing is particularly potent. The ratio of the area of the walls to the room volume dictates the concentration of formaldehyde in the air.
Solution: Use solid wood paneling, if you want to panel.
Corrective actions: Seal paneled area with a nontoxic sealant like Safe Seal. Air purification.

✛ CEILINGS

[ ] *Are any of your ceilings covered with asbestos tile?* Asbestos is sometimes found in acoustic tile, the kind you see in finished basements. Asbestos appears as a brownish material in acoustic tile.
Solution: Ensure that any acoustic tile you purchase is asbestos-free.
Corrective actions: If you suspect that your acoustic tile contains asbestos, ask an experienced remodeling contractor to verify it. Even though acoustic tile is easy to remove and exchange, you should still hire a professional abatement firm. If you do attempt to do it yourself, wear protective clothing, rent an asbestos-approved respirator, and contact your local hazardous waste group for instructions on proper disposal. Use a ventilator fan and air purification.

[ ] *Are any of your ceilings covered with acoustic tile?* Acoustic tile can contain fire retardants, solvents, and adhesives, including formaldehyde, which outgas and emit harmful vapors. Mineral fibers may also get airborne.
Solution: Use alternatives.
Corrective action: Air purification.

✛ LIGHTING

[ ] *Do you use fluorescent or incandescent bulbs to light your home?* Natural and artificial light are very different. Sunlight provides the whole spectrum of electromagnetic wavelengths in a specific mixture in which life has evolved on the planet. Artificial light only contains certain segments of the natural range in a very different mix. Incandescent bulbs provide light mainly in the red part of the spectrum while cool-white fluorescents emphasize the yellow-green portion. Many modern ailments have been associated with artificial lighting, including fatigue, depression, hyperactivity, diminished immunity, impaired fertility and deficiencies in calcium and vitamin D (necessary for healthy nerves, bones, and teeth). New discoveries link light directly to the production of hormones controlling our sexuality, fertility, growth, life span and the whole range of human emotions. In addition, linkages have been found between hormone levels and alcoholism, drug dependency and certain mental disorders. Since the average person spends between seventy-five to ninety percent of his time indoors, Dr. Wurtman (a MIT professor studying light impacts for twenty-five years) believes that most of us do not receive enough of the proper wavelengths of light for optimal health. He cautions against casual attitudes about light and health.
Solution: Spend more time outdoors and without sunglasses. Design your home to be well-lighted naturally. Skylights are excellent sources of daylight, allowing five times as much natural light as a sidewall window. Use window glass in homes, cars and sunglasses which allows the full spectrum of light to enter, including ultraviolet. (See *Product Sources*.)

⊣⊢ LIGHTING (Continued)

Corrective actions: Replace fluorescent and incandescent bulbs with full-spectrum bulbs. These bulbs simulate the natural spectrum in both the ultraviolet and visible light ranges. Dr. Wurtman believes that the spectrum of indoor lighting should deviate as little as possible from the spectrum occurring outside.

⊣⊢ ELECTRICITY

[ ] *Do you have small children and unprotected electric wall outlets?*
In 1993, there were nearly 3,000 injuries requiring emergency medical attention due to electrical shock from electric outlets. Most of these injuries resulted from small children inserting metal objects in these outlets.
Solution: There are now new outlets on the market with special barriers in each slot that stop even the most determined toddler from shoving everything from screw-driver blades to knives to paperclips into the slots. Yet power cords go in and out with just a firm shove or tug. There's nothing outwardly that reveals the special nature of these outlets. These would be especially important for child care facilities or for forgetful adults who just can't remember to reinsert those little outlet caps. Moreover, the plastic caps don't work if a child can get to the more vulnerable outlets by removing the caps or pulling out the electrical cords. (See *Shock-Patrol Outlet, Product Sources* in Appendix.)
Corrective Action: Although not as secure as the child-proof outlet, you can keep unused outlets covered with plastic caps designed to be difficult for the inquisitive child to remove. These can be found in most department or drug stores carrying baby supplies.

[ ] *Is your house located near high-voltage electric lines, radio or radar towers, microwave relay towers or an electricity-generating plant?*
Chaotic electromagnetic fields (EMFs) that enter your home radiate from these sources. Every living cell in the human body is regulated by electrical impulses controlled by the nervous system. Modern research shows EMFs interfere with the body's electrical field and affect our health in both gross and subtle ways. A study completed in 1986, for example, found a fivefold increase in childhood leukemia in homes located within fifty feet of primary high-current wires and twenty-five feet of primary, lower-current wires. Even if your home is not located close to large electric installations, you are still subjected to disordered EMFs from the wiring and appliances (especially from computers and TVs) in your home, car and office. Scientists do not yet completely understand the health effects of these fields, but many people report a sensitivity to them which affects mental clarity, anxiety levels, sleep patterns and moods.
Solution: Researchers advise a minimum safe distance of 200 feet from 50,000-volt lines and 500 feet from 350,000-volt lines. The intensity of the field does fall off with distance, but the quality of the field is the key to health. (See *EMFs and Coherence* in Appendix.) All other things being equal, locate as far from these sources as possible and design homes to minimize electricity use.
Corrective actions: Use an electromagnetic field coherence generator or other products to ameliorate EMFs. (See *Product Sources* in Appendix.) Unplug electrical devices when not in use. Switch off all noncritical electricity to the home at night at the fuse box.

⊣⊢ WATER

[ ] *Do you use unpurified water in your home?* Drinking and cooking is only one way that water pollutants enter the body. Over two thirds of your intake of water contaminants comes from inhalation or skin absorption from showering and bathing, operating dishwashers and washing machines and flushing toilets. Flushing the toilet without the lid down spreads thousands of bacteria through the air in a fine, breathable mist. Other contaminants also become airborne any time water is run or agitated, particularly hot water in the dishwasher or washing machine. These contaminants include chlorine, trihalomethanes (THMs), industrial pollutants, volatile organic chemicals (VOCs), arsenic, mercury, radon and radiation from human sources. (See *Toxic Substances and Health Effects* in Appendix.) The skin – a porous, permeable membrane – also absorbs toxins. Studies show the importance of skin absorption as a route of entry for many chemicals, including chlorine and VOCs.

Solution: Depending on your water problem, several whole-house water filtration systems can solve them. For example, a carbon system will remove radon and a UV system, bacteria. If you rent, a shower filter is the next best solution. (See *Water and Water Purification* and *Product Sources* in Appendix.)

Chapter 3   HOME FURNISHINGS

Our homes are now so full of plastics and other synthetic products that toxic smoke, not fire itself, is the cause of eighty percent of fire-related deaths. As with home building materials, synthetic materials should be avoided whenever possible and replaced with natural materials – solid wood, metal, glass and untreated cotton, wool and silk. Synthetic and chemically treated materials outgas toxic vapors that may cause acute or chronic illness and lead to degenerative diseases such as cancer.

✈ COMMERCIAL FURNITURE

[ ] *Is your home furnished in large part with new, commercially produced furniture, particularly pieces made from particleboard (in place of wood), synthetic stuffings and polyester fabrics?* Commercial furniture outgasses significant amounts of formaldehyde, particularly if it is made in whole or in part with particleboard or pressed wood and treated fabrics. Levels of formaldehyde can triple after a room is furnished. Be careful of overstuffed chairs and sofas as certain synthetic stuffings – foam rubber, polyurethane foams, and styrene foam chips – also outgas harmful vapors. Polyester fabric outgasses the most fumes. The material so overwhelmed the air filtration systems of NASA that nothing in space can be made of it.

[ ] *Is your home furnished with plastic furniture?* All plastics outgas harmful vapors, although this is a greater problem with soft rather than hard plastics used in the construction of furniture.

Solution: Buy furniture made of natural materials such as solid wood, metal, and glass, stuffed with cotton batting and covered with untreated cotton, wool, or silk. Be careful. Some furniture that may appear solid is made of pressed wood or plywood on the back, sides, bottoms and inside shelves. Particleboard can be very

convincingly veneered. If you are on a limited budget, used furniture will outgas less than new furniture. Re-upholster old furniture using natural stuffing and natural fibers. You can also buy solid wood, unfinished furniture and finish it yourself with Safe Seal.

Corrective actions: You can wash down particleboard or plastic furniture, even synthetic fabrics, with vinegar solution #1 about every six months. For fabrics, put the solution in a spray bottle and spray. In cases of extreme sensitivity, you may want to seal any such furniture with Safe Seal. Two coats should reduce fumes by eighty-five percent and four coats by ninety-five percent. Air purification.

✈ MATTRESSES

[ ] *Is your mattress or pillow made with polyurethane and treated with a flame retardant?* THPC, a commonly used flame retardant in mattresses, outgasses formaldehyde when wet. Polyurethane releases toluene disoyante.

Solution: Purchase untreated mattresses and pillows made from organically grown cotton and covered with unbleached, untreated cotton. Stuffing should be feathers (if you are not allergic) or cotton or wool batting.

Corrective action: Wash down your mattress with vinegar solution #2 and sprinkle with baking soda after dry.

12

⊣⊢ DRAPES AND BEDCLOTHES

[ ] *Do your sheets, blankets, bedspreads and drapes contain polyester or acrylic, or are they permanent press?* Polyester is a plastic that releases fumes and does not breathe or absorb perspiration well. Acrylic blankets may be worse. Acrylic outgases acrylonitrile, a toxic nerve gas. (The gas acrylonitrile is used mainly for two purposes – to make clothes and blankets and poison rats in grain silos.) Formaldehyde is used to make bedding and drapes "permanent press" and more durable. These items are usually more heavily treated than clothes and therefore outgas more. Fabrics treated with pesticides and dyed with synthetic dyes will also outgas harmful vapors. *Solution:* Bedding made of organically grown, untreated and unbleached cotton is best. Flannel is a good choice for sheets as it is normally untreated, but it is best to check with the manufacturer. Untreated wool or cotton is recommended for blankets. Look for fabrics with natural dyes. Recommended window treatments include untreated draperies of natural fibers, shades made of rice paper or cloth, wood shutters or metal blinds. *Corrective actions:* Polyester and formaldehyde treated materials can be made safer by washing with one cup of vinegar solution #1. In addition, put a cup of baking soda in the rinse cycle. If new, repeat the washings several times. If drapes are "dry clean only," spray them with the vinegar solution using a spray bottle. If new, hanging them outside for several days will rid them of most fumes. It is best to throw away anything made of acrylic. Air purification will assist.

⊣⊢ TOYS

[ ] *Do your children have many toys made from vinyl plastic containing polyvinylpyrrolidone (PVP) or polyvinyl chloride (PVC), or art or hobby supplies containing toxic materials?* Plastic toys may contain plasticizers and other components such as polyvinylpyrrolidone (PVP) or polyvinyl chloride (PVC) which can cause anything from minor irritation to degenerative diseases such as cancer. Children frequently put toys in their mouths, providing one route for entrance into the body. Plastic toys can also outgas harmful vapors into a child's room, especially if they are stored there. Be careful of glues and paints given to children, particularly if the labels contain a "danger" or "caution" warning. They may contain phenol, formaldehyde, or various plastics including PVP or PVC, or acrylonitrile, among other things. Even modeling clay contains free, crystalline silica or talc which may be inhaled, causing lung damage sometimes not apparent for fifteen or twenty years. *Solutions:* When possible, purchase dolls and toys from natural materials such as wood or cotton. Purchase natural glues and paints (See *Product Sources, Home Furnishings, Toys*) or use a nontoxic glue stick. Natural modeling beeswax is a good alternative to modeling clay.

Chapter 4   APPLIANCES

Appliances can be a source of pollutants and/or distribute pollutants throughout the home. These pollutants can vary from harmful gases, such as those produced by gas appliances and kerosene space heaters, to microwave radiation, to particulate matter spread by your vacuum or clothes dryer, to germs spread by a humidifier. In some cases, the best solution is to replace the item with safer alternatives. In others, proper venting may be sufficient.

⊶ GAS APPLIANCES

[ ] *Do you have a gas furnace, stove or hot water heater?* Improperly vented or maintained gas appliances are a source of air pollutants in your home. The most potentially dangerous is carbon monoxide (which is odorless and deadly in high concentrations). Other emissions include carbon dioxide, nitric oxide and nitrogen dioxide, sulfur dioxide, formaldehyde and tiny particulate matter. Unless the pilot light is electronic, it will burn and continuously release these pollutants into the home. Indoor air pollution levels can exceed those outdoors by five to ten times and can even rival Los Angeles on a bad day, particularly if the house is airtight (as in winter), a stove or oven is being used for supplemental heating, or the appliance is not vented or vented improperly. The health of the sick, shut-ins, the elderly and children is usually particularly affected. Two British studies linked gas cook stoves to higher incidence of respiratory illness in children aged five to eleven. A recent US study showed a large increase in childhood asthma, particularly in inner cities where gas stoves are often used to heat homes. *Solution:* Make sure that any gas appliance is properly vented to release waste gases outside. Seal off the areas containing gas heaters if possible. Use electronic lighting devices.

Corrective actions: Use gas stoves and ovens only for what they were intended. Clean and properly maintain any gas appliance. Buy carbon monoxide detectors. Air purification can help with some of the organic compounds released in the burning of natural gas, and negative ion generation or air filtration can assist with the particulate matter.

⊶ MICROWAVES

[ ] *Do you heat or cook food with a microwave?* Microwaves are now considered safe by most sources. If a product has a defective door, however, microwaves in concentrations that are potentially damaging to human tissue can leak out. *Consumer Reports* testing on microwaves found that leakage in normally functioning units was about one half milliwatts per square centimeter nearest the unit and declined quickly with distance from the machine. However, don't peer into the oven door to watch your food, or even stay in the room if you are pregnant. Microwaves are also heavy consumers of electricity and add significantly to electropollution. (See *Electricity.*) Those who do muscle testing as a diagnostic tool to determine suitability of a substance for use say that food cooked in microwaves is not optimal for health. (See *Muscle Testing* in Appendix.) *Solution:* Use alternatives to cook.

15

✈ VACUUMS AND DUSTING

[ ] *Do you have a portable vacuum?* Believe it or not, sweeping, dusting and vacuuming are very polluting. They may reduce the dust on furniture or the floor, but re-suspend small particles in the air that are likely to be inhaled (including pesticides, cleansers, pollen, mites, viruses, bacteria, and human skin scales). Vacuuming with a portable unit is particularly bad. Air is exhausted through the dust bag, throwing the more inhalable particles into the air. *Solutions:* Replace portable vacuums with those with well-designed micro-filtration, a water chamber, HEPA or a central vacuum system which carries dust into a main suction unit in the basement. (See *Product Sources.*) Seal the basement off from other living areas, especially if you also have a gas furnace and hot water heater there. (See *Gas Appliances.*) Avoid spray products for dusting and substitute nontoxic ones. (See *Aerosols.*) Use silicone treated dustcloths which hold the dust rather than resuspend it. *Corrective actions:* Negative ion generation and/or air filtration.

✈ HUMIDIFIERS

[ ] *Do you use a cool-mist humidifier, especially in the nursery?* Artificial heating can deplete the air of moisture, drying up the protective coating of the mucous membranes in the nose and throat and making us more susceptible to colds and flus. Cool-mist humidifiers, however, can breed and spread bacteria, viruses and spores, especially if they are not adequately cleaned and disinfected. *Solution:* Substitute an ultrasonic humidifier for the cool-mist models as these seem to spread fewer germs. (See *Product Sources, Appliances.*)

✈ CLOTHES DRYERS

[ ] *Do you have a gas dryer or a poorly vented dryer?* A gas dryer can expel the same pollutants into the air as your other gas appliances. (See *Gas Appliances.*) Electric or gas dryers that do not properly vent the hot air outside can put small respirable particles into the air. Clothes dryer belts made with asbestos can be the largest source of asbestos in the home. *Solution:* Make sure all dryers are properly vented to the outdoors. Hang clothes outdoors for the sun to dry. Buy a dryer belt made in the U.S. *Corrective actions:* Negative ion generation and/or air filtration.

✈ KEROSENE HEATERS

[ ] *Do you use a kerosene heater in winter?* Though more energy efficient than electric heaters, kerosene space heaters put out the same harmful pollutants as gas appliances, including carbon monoxide, carbon dioxide, nitrogen dioxide, sulfur dioxide and formaldehyde, and in very high concentrations. *Consumer Reports* showed that the concentration of gases which the heaters produced in a normally ventilated room reached levels high enough to be a serious health hazard to high-risk groups, including the elderly, children, pregnant women, asthmatics and people with cardiovascular disease, and could even threaten healthy adults.
Solution: Ask yourself why you need supplemental heat. If you do it to save on energy costs, consider buying an efficient heat-pump system and/or improving the insulation in your home. If you still want to heat only the rooms in use, make sure any kerosene or natural gas-fueled space heater is vented to the outside.

Chapter 5   CONSUMABLES

Persistent advertising has created multi-billion dollar markets for a plethora of household products, from cleaners to cosmetics to pet and car care formulas. The American public assumes these products to be safe. In the 1940s, only 1 billion pounds of synthetic chemicals per year were produced. Today, we manufacture some 500 billion pounds per year of over 70,000 different compounds for use in consumables. Many are unsafe or remain untested.

⌗ AEROSOLS

[ ] *Do you use commercial products dispensed by an aerosol spray?*
More than forty aerosol spray products can be found in the average American home, from hairsprays to oven and window cleaners, to deodorants and personal hygiene sprays. These sprays present a double danger: (i) They dispense propellants (from 4 - 25 grams in an average application) that remain airborne; and (ii) they release irritating or even toxic product ingredients in a fine, easily breathable mist. A few examples illustrate the possible dangers. Methylene chloride, a propellant, is carcinogenic and has been removed from most sprays except spray paint and thinners. Another propellant, nitrous oxide, has a deadening effect on the central nervous system. Cresol, a toxic chemical that attacks the central nervous system, kidneys, liver, spleen and pancreas, is commonly found in disinfectants. Aluminum chlorohydrate or other aluminum compounds are often the active ingredients in deodorants. Aluminum is a toxic element potentially connected to Alzheimer's.
Solution: Use alternatives to aerosols whenever possible. (See discussions below of *Cleaners* and *Personal Care Products,* and *Recipes* in the Appendix.)

⌗ CLEANERS

[ ] *Do you have a house full of commercial cleaning products?*
The active ingredient of most commercial cleaners is usually highly caustic, toxic, irritating or even fatal. For example, sodium hypochlorite, sodium hydroxide, and phosphoric acid – main ingredients in many bleaches, disinfectants, and bathroom, drain, and oven cleaners – severely burn the eyes, skin, internal organs and lungs, and can be fatal if enough is ingested. In addition, they probably contain one or more of the following: petroleum distillates, artificial coal tar dyes and artificial fragrances. Some active ingredients, and even the petroleum-based solvents, may be carcinogenic, mutagenic (causing mutations) or tetratogenic (causing birth defects). (See *Toxic Substances* in Appendix.) For example, most all-purpose cleaners contain DEA and TEA (See *Personal Care Products*) which react with nitrates to form carcinogenic nitrosamines that readily penetrate the skin. Butyl cellosolve, a neurotoxin, is commonly found in glass and window cleaners, carpet cleaners and all-purpose cleaners. Arsenic, lead or other impurities in artificial dyes can be absorbed through the skin and cause cancer or illness.
Solution: Buy products known to be safe or make your own. (See *Recipes* and *Product Sources* in Appendix.)

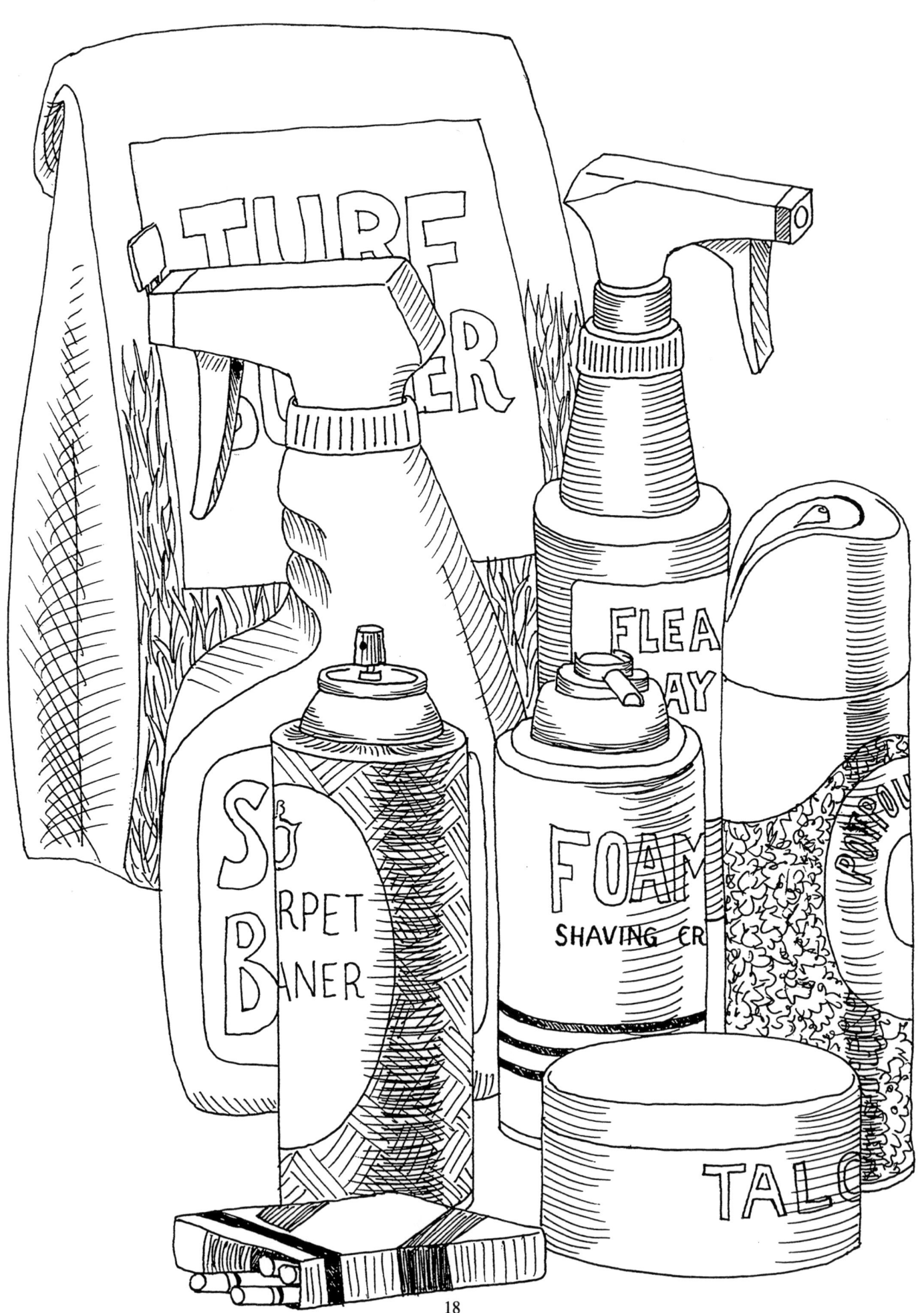

18

✈ PERSONAL CARE PRODUCTS

[ ] *Do your personal care products contain synthetic chemicals?*
Most people do not realize that the skin is the body's largest organ and is very porous. Anything put on the skin is readily absorbed into the body, so the quality of your personal care products is doubly important. Be cautious. Products labeled "natural" frequently aren't and contain many ingredients that don't belong on your face, hair or skin. For instance, the National Institute of Occupational Safety and Health found that 884 of the chemicals available for use in cosmetics have been reported to the government as toxic substances. (See Appendix, *Labeling and Regulation.*) Substances regularly used in lotions, soaps, shampoos, toothpastes, hair dyes, and cosmetics, among many others, can cause anything from minor skin irritations, allergies and sun sensitivity to more serious acute or chronic problems like cancer, asthma, eczema, depression, nausea, spaciness, lethargy, hyperactivity, irritability or hypersensitivities. Commonly used chemicals which should be avoided include: **1. Artificial fragrances.** These are one of the two leading causes of allergy, irritation or more serious problems from cosmetics. The word "fragrance" on the label usually means a synthetic chemical, not an essential oil. It could be any mixture of hundreds of ingredients, including hazardous chemicals such as toluene, methylene chloride or ethyl alcohol.
2. Preservatives. These are the second leading cause of contact dermatitis. The parabens (butyl, ethyl, methyl, propyl) and preservatives that contain or may release formaldehyde are the most important ones to avoid.

3. Synthetic detergents. These do the same kind of harm to our environment as detergent laundry products and are far less gentle on the skin and hair than natural soaps made from vegetable oils and sea salt. Sodium lauryl sulfate is common in shampoos.
4. Wetting agents. Agents such as DEA (diethanolamine) or TEA (triethanolamine) are often combined with detergents and can form carcinogenic nitrosamines during formulation or even as they sit on the shelf. Also avoid ethoxylated wetting agents that may be contaminated with 1,4-dioxane, a carcinogen.
5. Mineral oil and its derivatives. When used as moisturers, they interfere with the body's own moisturizing mechanism and make the skin prone to sun damage. Common mineral oil derivatives include petrolatum, glycerol stearate and propylene glycol.
6. Artificial colors. Indicated by D&C and F,D&C labeling, these are sometimes carcinogenic, even when applied to the skin.
7. Lanolin. Although natural, it can be contaminated with pesticides.
8. Cosmetic talc and crystalline silica. The hazard of these carcinogens arises primarily from inhalation. Some studies indicate using talc in the genital area increases the risk of ovarian cancer.
9. Chemical hair dyes. Permanent and semi-permanent dyes deserve a special mention because of their association with an increased risk of certain types of cancer, including breast cancer. The use of hair-coloring products accounts for twenty percent of all non-Hodgkin's lymphoma cases in women.
Solution: Buy products known to be free of these chemicals. (See *List of Cosmetic Ingredients to Avoid* and *Product Sources* in Appendix.)

+ PAPER PRODUCTS

[ ] *Do you consume paper products bleached with chlorine or containing artificial fragrances or dyes?*
Paper products that have been bleached by chlorine may contain dioxins, one of the most dangerous synthetic compounds ever produced, causing skin disorders, cancer and mutations. Dioxin may be absorbed through your skin from these products (not to mention that their manufacture contaminates the environment). It is important to verify that any paper product you use hygenically (facial tissues, toilet paper, napkins, tampons and feminine pads, and diapers) are not bleached with dioxin producing processes. Artificial fragrances and dyes in these products have also been linked to itching, burning, cramps, and even herpes-type flare-ups. (See *Personal Care Products*, and *Toxic Substances* in Appendix.)
Solution: Switch to unscented, white paper products that have been bleached with hydrogen peroxide, an alternative to chlorine processes. Also use 100% post-consumer recycled paper products or, where possible, organically grown, unbleached cotton alternatives. This also helps conserve forest resources.

+ AIR FRESHENERS

[ ] *Do you use aerosol or other artificial or perfumed air fresheners?*
Most air fresheners do not purify the air, but only add more pollutants in an attempt to cover up offensive odors. They work in one of three ways: (i) coating your nasal passages with an oil film; (ii) using a nerve-deadening agent to interfere with your ability to smell; or (iii) covering up one odor with another. (See also *Aerosols*.)

Solution: Keep things clean and open windows. Place open boxes of baking soda in enclosed areas, and place fragrant flowers and herbs or bowls of petals in several rooms. Sunshine is a natural disinfectant. Small bags of silica gel (obtained from camara stores) help keep moldy areas dry. Certain plants perform a natural air purification function. For example, spider plants remove carbon monoxide and formaldehyde; English ivy eliminates benzene; and potted mum removes trichloro-ethylene. Depending on the sources of odor and your health status, quality air purification equipment may be necessary.
(See *Product Sources* and *Recipes* in Appendix.)

+ TOBACCO

[ ] *Does someone in your home smoke?* Tobacco smoke is a complex mixture of more than 3,000 chemicals, some of which are additives, and many of which are toxic, carcinogenic or even radioactive. Tobacco smoke aggravates the health hazards of nearly every indoor pollutant and poses health threats to anyone who breathes it. Studies show that those passively exposed to cigarette smoke have two to four times the risk of incurring lung cancer. Children whose parents smoke are at a higher risk of pneumonia, bronchitis and other respiratory tract infections, and of dying of leukemia. Moreover, once indoor air is polluted, it takes awhile for it to be free of noxious vapors. Smoke pollutants cling to walls, draperies, clothing and furnishings where they are released slowly for weeks.
Solution: The best solution is not to smoke or allow others to smoke in your healthy home. In the meantime, air purification and negative ion generation can help purify the air.

✈ CAR CARE

[ ] *Do you use commercial products for cleaning and protecting the interior and exterior of your car?*
Many commercial products used to maintain your car's appearance are flammable or contain chemicals irritating to the eyes, skin or respiratory tract. Some others pose small to large cancer risks and cause cumulative damage to the nervous system or reproductive system. Be particularly wary of interior cleaners that contain formaldehyde or you will be breathing this chemical every time you drive your car. Washes and waxes may contain artificial colors or other types of carcinogens. (See *Cleaners.*)
Solutions: Safe alternatives to these types of products are available. To wash your car, simply pour 1/4 to 1/2 cup of a recommended liquid soap or dishwashing detergent into a bucket of warm water. For car interiors, use 1 to 2 teaspoons of washing soda dissolved in 1 cup of boiling water. Shine chrome with 1/4 cup of baking soda saturated with water to make a paste, or a cotton cloth saturated with lemon juice or vinegar. Tar can be removed with boiled linseed oil or a plant-based paint thinner. No alternatives to car wax are available, but some products are safer than others. (*See Product Sources.*)
Corrective Actions: If you must use commercial products, apply them in an open or well-ventilated area and wear impermeable gloves.
[ ] *Does your anti-freeze contain ethylene glycol?*
The sweet taste of ethylene-glycol makes it tempting to children, pets and wildlife. Tens of thousands of animals die annually, accounting for half of all poisoning deaths of pets. Ethylene glycol also poses neurotoxic and reproductive risks.

Solution: Use an antifreeze formulated with propylene glycol, a milder ingredient proven much safer and longer lasting in automotive use. It appears to last two to three times longer than ethylene glycol. (See *Product Sources* and *Disposal of Hazardous Wastes* in Appendix.)
[ ] *Do you know how to properly dispose of automotive products?*
Most automotive products, particularly those for engine maintenance, pose dangers to human health and the environment. Unfortunately, except in the case of antifreeze mentioned above, no commercially available alternative to the use of these products exists except minimizing use of your car. Walk, bicycle or use public transportation.
General Rules: If you do your own maintenance, follow these rules:
(i) Never pour anti-freeze, used motor oil or other fluids on the ground.
(ii) Gasoline, antifreeze and most degreasers with hazard on the label should be taken to a licensed hazardous-waste disposal facility.
(iii) Recycle car batteries at a local battery shop and used motor oil at an oil recycling center or gas station collection point. Transmission and brake fluid, diesel fuel and kerosene can be added to oil and recycled if the treatment facility can handle these fluids. Contact the county extension office.
General Car Information
Driving today's petroleum-driven car has an extremely high environmental impact. (Prototype cars have been built that run on other fluids, even water.) In the U.S., cars account for about half of all oil consumption and about a third of all the greenhouse gases emitted every year. Oil production and refining is one of our more polluting activities. (See *General Environmental Issues* in Appendix.)

⊱ PEST CONTROL

[ ] *Do you use commercial products to control insect pests in the home?* Despite clear evidence that home pesticide use puts you and your children at risk, nine out of ten American households use pesticides, accounting for thirty percent of the $5 billion annual worldwide pesticide sales. The U.S. now manufactures more than 200 billion pounds of active pesticide ingredients, representing over 1,400 different active compounds. The great majority of these have never been adequately assessed for their carcinogenic, mutagenic, neurotoxic or other properties.

Pesticides are the number two cause of household poisonings in the U.S. As Rachael Carson states, "pesticides (sic) have immense power not merely to poison but to enter into the most vital processes of the body and change them in sinister and often deadly ways." Some pesticides affect the central nervous system, causing depression, memory loss or impaired thinking, even when small amounts are inhaled. This is alarming as most pesticides are formulated to act over long periods of time and can remain actively airborne for weeks or months. Other pesticides are stored in body fat, accumulating over time to toxic levels to cause cancer, damage to vital organs, or reproductive problems. For example, childhood leukemia is four times higher than normal in homes where pesticides are used during pregnancy. Use of pest-strips and flea collars with dichlorvos (DDVP) pose a high cancer risk even to pets.
Solution: Natural pest control starts with the elimination of entrances to your home and favorable living conditions for pests. Fill cracks and holes and buy screens for your

windows and doors. Take away their food and water supply by keeping living areas clean, storing food in impenetrable containers, and repairing leaky faucets. Empty the garbage frequently and get rid of clutter where pests can hide.

A "pest" can be only a matter of our perception. Some organisms considered pests are truly beneficial. Spiders and centipedes eat flies, for instance. The insects considered hazardous to health by the World Health Organization (WHO) include fleas, bedbugs, lice, mites, scorpions and some spiders. Cockroaches are not included. The scientific literature does not report a single case of cockroaches transmitting human disease.

Corrective Actions: Most "pest" problems can be easily solved with simple nontoxic measures such as a vacuum cleaner, a sprinkle of cayenne pepper or borax (with sugar) or a few sprays with soapy water or mint tea. Some simple home remedies follow.
Ants: Sprinkle borax or powdered chili pepper where ants come in, and spray ants and ant trails with strong mint tea or an organic soap.
Beetles and Weevils: Store grain products in the refrigerator or seal in tight glass containers with a bay leaf.
Cockroaches and Silverfish: Spread around infested areas equal parts baking soda and sugar or a mixture of 1 part trisodium phosphate, 6 parts borax, 4 parts sugar and 8 parts flour.
Flies: Catch them by spreading a thin layer of honey onto bright yellow paper or use a swatter.
Moths: Clean clothes and lay them in the sun or run them through a dryer to kill moth eggs before storing; store in airtight containers with cedar chips.
Spider Mites: Mix a natural spray for plants from four cups wheat flour, a half cup buttermilk and five gallons of water; fertilize plants with compost.

PEST CONTROL (Continued)

[ ] *Do you use a commercial insect repellent containing DEET?*
The British medical journal, *Lancet*, and *Consumer Reports* report DEET to be neurotoxic, particularly to children. It absorbs through the skin and at high enough dosages can cause slurred speech and difficulty walking, tremors and even death.
Solution: Use mosquito netting at night and wear protective clothing. Use alternatives to commercial repellents – vinegar or natural products. (See *Product Sources, Pest Control* in Appendix.)

[ ] *Do you have a termite bond with a professional termite exterminator?*
Most professional exterminators treat termites by spraying long-lasting and potent pesticides in the crawlspace or basement and in the soil around foundations. These can persist in the air one year later, even in properly treated homes, and for decades underground. Chlorpyrifos (commonly known as Dursban, Lorsban, or Pyrinex) is now the most popular termiticide, replacing chlordane as the favorite until it was banned in 1987.

Chlordane was used so lavishly for twenty years that it probably still contaminates many older homes. The chemical attacks the central nervous system, affecting the brain. Symptoms of persistent exposure include nausea, headaches, anxiety, irritability and exhaustion. Even when properly applied, chlordane seeps into the home, polluting the air for years. When accidently pumped into basements or heating ducts, the chemical gets absorbed by floors and walls which then outgas toxic vapors. Many homes have been abandoned and condemned due to chlordane poisoning. If you suspect that your home has been affected, you can have it tested for pesticides.

Chlorpyrifos, chlordane's most popular replacement, has not been properly tested, according to the Consumers Union. It is known to be extremely toxic to birds and fish. The *Safe Shopper's Bible* lists chlorpyrifos as both a reproductive toxin and a neurotoxin, and warns against the use of it, as well as Vikane and methyl bromide.
Solution: The best solution is to prevent termites by making your home unattractive to them. Termites need warmth and moisture, so keep the area under your house cold and dry. Make sure soil under and around your home is well drained. Desiccating dusts can be injected into the walls. Subterranean barriers between the earth and your home's foundation can be used. Have your contractor install copper or galvanized steel termite shields. Apply barrier sand before new construction or around an existing structure. Wooden floor joists should be twenty-four inches above the ground and exterior wood at least six inches off the ground. Any wood that is in contact with the ground should be heart redwood or heart tidewater red cypress. Remove any scrap wood, firewood and stumps or other food sources from around the home.
Corrective Actions: If you have termites, cutting out and replacing the infested wood is often sufficient. Other alternatives include nematodes, a nontoxic biological approach; heat treatments (apply a heat lamp at 140o for 10 minutes); cold treatments using liquid nitrogen; and high voltage electrical currents. A chemical alternative is Timbor, by U.S. Borax, mixed with soap and applied with a foaming machine. Most of these would require a professional experienced in these alternatives. (See *Further Resources, Pesticides* in Appendix.)

⊬ LAWN/GARDEN CARE

[ ] *Do you use artificial fertilizers or pesticides or a lawn care company that does?* Pesticides, which include insecticides, herbicides, and fungicides, are a group of poisons used for killing unwanted organisms, but often harm beneficial nontargeted animals, including human beings. For example, children whose parents use pesticides in their homes or gardens before their birth have a risk four times higher than normal for child-hood leukemia. Childhood brain cancer is associated with the use of diazinon and carbaryl (Sevin) in the garden and with herbicides used to control weeds. Common pesticides are the second leading cause of household poisonings in the United States. Seventy percent of these incidents involve children.

Artificial fertilizers are also detrimental, not only to humans, but also to the soil, air and water. Artificial fertilizers are made up of nitrogen, phosphorous and potassium. Nitrogen, as anhydrous ammonia, can burn the skin and lungs and reduces beneficial micro-organisms in the soil. Potassium (as potassium chloride) produces chlorine gas and hypochlorous acid, which also kill soil bacteria. Both produce gases that harm the atmosphere. Chlorine and nitrogen gases contribute to ozone depletion, and nitrogen contributes to acid rain and global warming. The predominant use of these three minerals, while giving plants an impressive jolt of growth, does nothing to address the other thirteen essential elements (and fifty-three other micro-nutrients detected in plant life) needed for good nutrition. Most commercially produced plant foods are deficient in trace elements we need to produce proteins and amino acids

in our own bodies. For example, the widespread use of potassium has caused a drop in magnesium in soils and in our diets. Magnesium is important to our ability to relax and handle stress.

Pesticides and fertilizers leach into ground water and drain into sewers and local water supplies. They also evaporate into the air during and even after application, especially when sprayed from a nozzle from large tanker trucks by lawn care companies. Ironically, weeds and pests both appear to be nature's way to correct deficiencies and eliminate sick and dying plants. Neither cause problems when soil is healthy and plants have not been genetically inbred. Weeds create conditions that return missing elements back to the soil. Clover, for instance, is an indicator of low nitrogen and has root nodules that convert nitrogen to a form usable to plants. Insects and fungus are our garbage collectors, and are normally only attracted to nutrient-deficient, weakened and diseased plants. Healthy plants in healthy soils keep pests away with their own natural pesticides. *Solutions:* If you have a choice, consider not growing a lawn. Leave the natural ecosystem around your home or landscape for wildlife. If you already have a lawn, have soil tests done by your local extension service every few years and use only natural fertilizers and soil amendments. Learn to compost. Compost is nature's most nearly perfect soil conditioner and plant food. Aerate twice a year, de-thatch in late spring or early summer and reseed, preferably with local varieties. Mow only when necessary, mow high and leave the grass clippings. Cut dandelions at the root with a fishtail weeder. Release natural predators. (See *Natural Lawn and Garden Care* in Appendix.)

✠ PET CARE

[ ] *Do you use commercial cat litter?*
Many cat litter products contain crystalline silica, a product not only irritating to the lungs, but also carcinogenic, although not a major hazard. Cat litter can also contain strong deodorants and artificial fragrances. Cat feces transmits a disease called toxoplasmosis, to which pregnant women are very susceptible.
Solutions: If appropriate, train your cats to go outside. If you use a cat box, use litter without crystalline silica or artificial fragrances. Wash hands thoroughly after handling cat litter and do not handle it if pregnant.

[ ] *Is your pet bed made with polyester or stuffed with styrofoam?*
Polyester is not any better for your pet than for you. (See *Mattresses, Drapes and Bedclothes.*)
Solution: Use pet beds made and stuffed with natural materials, or make your own from burlap or heavy cotton. Fill it with cedar shavings and cotton.

[ ] *Do you use standard chemical flea or tick collars, dips or sprays?*
These products pose hazards to both pets and their owners. The active ingredients include pesticides like diazinon and carbaryl, nerve poisons that may cause long term health problems. Carbaryl can also cause birth defects in dogs. Signs of nervous system poisoning include salivation, muscle tremors, vomiting, staggering, even death. Dusting can cause headaches and nausea. Foggers and bombs expose everyone in the house to unhealthy chemicals. Moreover, strong poisons are only temporary measures if the underlying problem remains. Flea infestations are usually signs that your pet suffers from poor nutrition and poor skin condition.

Solutions: Feed your pet a healthy diet. (See *Pet Food.*) Give your pet daily tablets of brewer's yeast and/or one third or less of a teaspoon of blue green algae powder per pound of **food**. For skin conditions, make a lemon skin tonic. Thinly slice one whole lemon, add it to one pint of near boiling water. Steep the mixture overnight and sponge the solution on your pet's skin daily. Shampooing your pet can drown and knock off fleas. Use a natural soap or an insecticidal soap that uses nontoxic fatty acid salts and potassium oleate as active ingredients. Use a flea comb to capture the pests, then drop them into soapy water and flush them down the toilet. To rid your home of fleas, dust with borax or brewer's yeast, then vacuum thoroughly and wash the pet's bedding. Afterwards, remove and seal the vacuum bag and put it in the sun or the freezer to kill the fleas. To keep them away in the yard, sprinkle flea areas with diatomaceous earth, an abrasive crystal that cuts insects. Plant oils, such as pennyroyal, lavender, citronella, eucalyptus and rosemary oils, are also good flea repellents, but can cause contact dermatitis in some dogs. Use these in herbal flea collars, a half teaspoon added to your pet shampoo, in 1:5 or 1:15 dilutions with water rubbed on the fur, or sprinkled over rock salt (two ounces to two to three quarts) to spread in the house. Pennyroyal is toxic to cats. Also avoid using pennyroyal if you or your pets are pregnant. In severe cases, try natural pyrethrums with piperonyl butoxide, or shampoos with these ingredients, or bake the fleas out of your house. After taking all living plants and animals out, turn up the heat to the highest setting and leave the house for a day. (See *Consumables – Nutritional Products* and *Pet Care* under *Product Sources* in Appendix.)

Chapter 6   FOOD

Good food, next to fresh air, sunshine and exercise, is the most important factor in general health. We literally are what we eat. Sadly, modern agriculture has depleted the soil, giving us malnourished plants, which lack many micro-nutrients necessary for full vitality. In addition, we are now getting more of what we do not need such as pesticides, synthetic chemicals and drugs. Cultures who still eat whole, unrefined foods have a remarkable lack of chronic conditions, like cancer and heart disease.

✦ SYNTHETIC CHEMICALS

[ ] *Do you eat a large proportion of processed foods and meat containing food additives and preservatives?*
The average American eats six pounds of synthetic compounds, representing five thousand different chemicals, in his or her food each year. Most of the thousands of food additives have never been completely assessed for health hazards. We also don't know how they interact, though we normally consume many together. One of the few studies is alarming. Rats given *one* of three common food additives (sodium cyclamate, Red Dye No. 2 and polyxyethelene sorbitan monostearate) showed no change. When given two, they developed diarrhea, lost their hair and stopped growing. When given all three, they lost weight and died in three weeks.

Coal tar food colors, BHA and potassium bromate are carcinogens of greatest concern. In addition, nitrite preservatives found in luncheon and cured meats form carcinogenic nitrosamines in the food or in your stomach. Other synthetic chemicals in food include sulfa drugs and other antibiotics given to livestock and farm-raised fish. Illegal residues in meat, fish and poultry can be carcinogenic or cause birth defects. Growth hormones given to ninety percent of cattle are also carcinogenic.

Solution: Eat fresh, organically grown fruits and vegetables, whole grains and nuts as much as possible. If you want convenience food, look for "natural" convenience foods that are free from artificial colors, flavors and preservatives. If you choose to eat meat, buy organically raised meat. Choose wild, deep water ocean fish to farm raised fish whenever possible.

✦ PESTICIDES

[ ] *Do you know which fruits and vegetables have fewer pesticides or do you buy organic produce?*
A 1987 EPA report contends that pesticide contamination of food is the number one health problem in the U.S. Ninety percent of fungicides, fifty percent of herbicides, and thirty percent of insecticides cause cancer in lab animals. More frightening, two 1993 studies conclude that children and infants are likely to develop future cancers because of exposure to carcinogenic pesticides in their diet. Another 1993 study showed women with the highest levels of DDT had a four times greater risk of breast cancer than those with the least exposure. One in three Americans now contracts cancer, up from one in four in 1950. Pesticides not only contaminate food, but are reaching high concentrations in our bodies, air, water and soils. Yet pesticide usage continues to rise as effectiveness declines.

✈ PESTICIDES (Continued)

The products that put you most at risk for pesticide contamination include non-organic meats and dairy products, certain fish and seafood and certain fruits and vegetables (if not organic). (See adjacent column.) Other products at risk for pesticide residues include applesauce, baking mixes, flour and bread, beans, cereal, chips, coffees and teas, fish, herbs and spices, jams, jellies and preserves, juices, ketchup, nuts and peanut butter, olives, oils and salad dressings, pasta and pasta sauces, salsa, soup, soy sauce and processed cane sugar. *Solution:* Buy foods and produce that have been certified to be organically grown. "Certified" means that an independent organization has verified that the producer grows, processes, packages and transports food using methods that meet certain criteria and standards. Just because it is in a health food store doesn't mean that it is organically grown. Ask questions of your produce buyer. Look for the certifying groups' logos. (See *Internationally Recognized Certifying Groups.*) Can you be sure that certified organic produce really is better? Several analyses, at least in California and Washington, have concluded that "the amount and concentration of pesticides detected in organic produce is significantly lower than for conventional produce." An added bonus is that organic produce is also free from artificial colors, artificial preservatives and drugs. Studies have also found organically grown food to be higher in nutrition. Conversely, a food may still be organic even if it is not certified. In particular, locally grown produce sold at farmers' markets may be organic, but you need to verify this with the growers.

Presuming Non-Organic Produce from Conventional Stores The Twelve Most Contaminated Fruits and Vegetables:

Strawberries
Bell Peppers
Spinach
Cherries
Peaches
Cantaloupe (Mexican)
Celery
Apples
Apricots
Green Beans
Grapes (Chilean)
Cucumbers

The Twelve Least Contaminated Fruits and Vegetables:

Avocados
Corn
Onions
Sweet Potatoes
Cauliflower
Brussel sprouts
Grapes (U.S.)
Bananas
Plums
Green onions
Watermelon
Broccoli

Source: Center for Science in the Public Interest.

Some Internationally Recognized Certifying Groups:

Biodynamic
California Certified Organic Farmers (CCOF)
Farm Verified Organic (FVO)
Florida Organic Growers and Consumers
Natural Organic Farmers Association
Organic Foods Production Association of NA
Organic Growers and Buyers Association
Virginia Association of Biological Farmers
Wisconsin Natural Foods Associates

Source: The Safe Shopper's Bible (1995)

28

⚊ IRRADIATION

[ ] *Do you eat foods that have been irradiated?* Radiation entails exposing foods to massive doses of radiation (roughly equal to 10 million medical X-rays) to kill bacteria, molds and insect larvae. The notion of irradiating foods to prevent spoilage was first proposed in the 1950s by the Atomic Energy Commission in their effort to find uses for radioactive waste from nuclear weapons. Industry and the FDA proclaim the safety of irradiated food, but research into the long term effects by the Department of Energy and the Pentagon have never been made available. While food itself is not made radioactive, its chemical structure is altered. Results include the formation of benzene and other by-products. Test animals fed whole, irradiated food show chromosomal and reproductive damage. Indian children fed freshly irradiated wheat also showed chromosomal damage. Moreover, workers in the irradiating plants have been exposed to high levels of radiation in accidents. *Solution:* Boycott the purchase of irradiated food and seeds. Foods likely to have been irradiated include fruits, vegetables, pork, herbs, spices and teas. Whole, irradiated foods should display the international logo of a flower in a circle and the words, "treated by irradiation." Spices and teas are more difficult to distinguish. Generally, imported spices and those used in processed foods have been irradiated. McCormick, a leading spice company, does not irradiate food and states so on their labels.

⚊ COOKWARE AND DISHES

[ ] *Do you use aluminum or stainless steel cookware, or pottery that was imported or made before 1970?*

Cooking or storing food with aluminum can cause above-normal levels to build up in your body. This includes the use of disposable aluminum pans or foil. Even stainless steel cookware which contains eighteen percent chromium and eight percent nickle leaches small amounts of these metals into your food. Under-glazed ceramic pottery can leach lead into food and cause lead poisoning. *Solution:* The cookware that is safest is made of glass or ceramic. Stainless steel is also recommended if it is of the highest quality and engineered so as to minimize extra water needs and cooking time. Use wood or hard plastic, not metal, to stir foods. The inside liner of a plastic thermos should be glass. Ceramic cookware or dishes from Italy, India, China, Mexico or Hong Kong are likely to leach lead. Use questionable ceramic or pottery only for decoration.

⚊ PACKAGING AND STORAGE

[ ] *Do you purchase or store foods in styrofoam, cling wrap, plastic wrap or plastic (PET) containers?* Polystyrene foam (styrofoam) is one of the worst forms of packaging for food. If you eat or drink food contained in styrofoam, you ingest styrene, a neurotoxin and suspected carcinogen. Cling film used in packaging cheese, meat, fish and some fruits and vegetables contains carcinogens which migrate into food. Plastic wrap contains the carcinogen vinylidene chloride. Microwave packaging also has poorly studied chemicals that migrate into foods, especially during microwaving. Polyethylene terephthalate (PET), used in plastic beverage containers, can release plasticizers at elevated temperatures. Bisphenol-A, which leaches from five-gallon jugs, is an estrogen mimic.

Solution: Buy fresh vegetables and fruits without cling wrap packaging. Have fresh meat, poultry and fish wrapped in butcher's paper or waxed paper. When buying dry bulk foods, make sure the bins use food grade material with no glue joints. Avoid microwave foods in general. (See *Microwaves* under *Appliances.*) Use wax paper, recycled glass jars or Pyrex glass containers for food storage. Hard plastic (polypropylene) containers are acceptable if they are first washed in vinegar solution #1. (See *Recipes* in Appendix.) Cellophane made from plant cellulose is also acceptable. As much as possible, buy beverages contained in glass containers.

⊣⊢ BEVERAGES

[ ] *Do you consume a large amount of coffee or decaffeinated coffee?*
Heavy coffee consumption (seven cups or more daily) is a possible cause of urinary bladder cancer. Heavy consumption can also complicate pregnancy. Spontaneous abortions, stillbirths and premature births are associated with high consumption by the father; while increased incidence of birth defects are associated with high consumption by the mother. Many brands of decaffeinated coffee use methylene chloride, a carcinogen, in the decaffeination process.
Solution: Switch to herbal teas or coffee decaffeinated using the Swiss water process. (If you are pregnant, consult with a health practitioner about your herbal tea.) For more energy, exercise and take nutritional or other supplements. (See *Energy and Nutritional Supplements.*)
[ ] *Do you drink bottled water?*
If you are relying on bottled water for health-related reasons, beware. The quality of bottled water is regulated no more strictly than tap water, and

quality can vary tremendously. Bottled water from deep springs is of course better than water near a chemical or nuclear plant. Hardly any place on earth is free from pollution, however. Even the polar ice caps have PCBs and DDT, and spring water must be filtered. A wide range of pollutants have been found in bottled water, including cancer-causing radiation and trihalomethanes (THMs), a by-product of chlorination. An EPA survey of twenty-five bottling plants found sanitary deficiencies in all facilities, and eight percent of the water sampled contained coliform bacteria, an indicator of disease-causing bacteria. Even where water quality is good going into the bottle, plastic from the container can leach into the bottle, especially if heated (See *Packaging and Storage*), and bacteria can grow in water sitting too long on the shelf.
Solution: If unsure, filter your water in your home. (See *Water.*)
[ ] *Do you drink more than one alcoholic beverage a night?*
While studies show two to four drinks a week can lower death rates from heart disease, two or more drinks a day lead to sixty-three percent higher than average death rates. At higher consumption, the lower death rate from heart disease is offset by higher death rates from cancer. This is not surprising. Certain alcoholic beverages contain a wide range of carcinogens, including asbestos, lead, pesticides and urethane. Urethane is one of the most dangerous of the chemicals, causing breast cancer in 100 percent of experimental animals and in single doses. Asbestos and urethane are found in a range of beverages, but lead and pesticides are generally in wine. (See *The Food Shopper's Bible.*)
Solution: Drink less alcohol and find other means of stress relief.

✠ BEVERAGES (Continued)

[ ] *Do you currently drink many sodas?* Despite the fact that soft drinks are socially prevalent and acceptable, they can be as addicting as coffee and alcohol. According to *Young Again* by John Thomas, their impacts are experienced over a long period of time and the list of conditions associated with their long term consumption is staggering! (See *Aging and Modern Living* in Appendix.) The sugar in soft drinks steals the body's mineral ions. Aspartame has other impacts. (See the next section on *Sweeteners.*) According to Thomas, sodas are hard on the kidneys and create imbalance in the body's pH due to their sodium content and traces of phosphoric acid. Thomas states sodas contain aluminum ions from the can as well. Aluminum is potentially associated with Alzheimer's and other neurological conditions.
Solution: Detoxify the body to end cravings. Drink plenty of water for thirst. For snack beverages, make your own fruit or vegetable juices from organic produce.

✠ SALT

[ ] *Do you currently consume large quantities of table salt or processed foods high in sodium?* Sodium imposes a high stress on the body. Normally, sodium is supposed to remain outside the cells of our bodies in the extracellular fluid. When the body has insufficient potassium, and excesses of sodium, it replaces the potassium inside the cells with sodium. Both ions have a positive charge and both are electrolytes of about the same size. However, their impact inside the cell is quite different. Once inside the cell, sodium interferes with the functions of the mitochondria and other celluar machinery. The mitochondria are the key powerhouses for our cells. Shutting them down is like cutting off the power and eventually the cell dies. High salt intake will eventually lead to all kinds of degenerative diseases.
Solution: Replace salt with Bragg's Liquid Aminos, liquid sea minerals, and/or Celtic Sea Salt. Substitute processed foods high in sodium with home grown or organic fruits and vegetables. Good food from healthy seed stock, grown on rich, living soils tastes good without salt!

✠ OILS

[ ] *Do you use soy or canola oil?* According to John Thomas (above), soy and canola oils are industrial oils; they thicken and become gummy when heated, especially after they cool. Olive oil, butter or lard become thinner. Soybean oil contains a toxic biochemical, *phytohemaglutinin*, that slows blood circulation, clots blood, closes off capillaries in the posterior eye, ears and scalp (hence baldness), negatively influences the central and peripheral nervous systems and magnifies problems with Rouleau ('sticky blood'). Soy oil also reacts with dissolved gases and minerals in the blood like chlorine, chloramine, and flourine to form plaque that adheres to arteries that service the heart and brain. Canola oil is another industrial oil used as a lubricant, in soap and synthetic rubber, and as an illuminant for slick, color magazine pages. It also causes agglutination of the red blood cells and can lead to loss of vision, deterioration of the central and peripheral nervous system, constipation and other problems.
Solution: Stick with olive, sesame, sunflower and flax oils. Avoid corn, peanut, safflower and cotton seed oil.

✈ MEAT

[ ] *Does a large proportion of your diet consist of red meat?* There are many reasons to cut down your consumption of meat. The most important are the linkage of high meat consumption to cancer and heart disease. More Americans die each year from cardiovascular – heart and blood vessel – diseases each year than from all other causes combined. According to research, four things lead to the clogging of arteries and heart attacks and strokes: a high level of one type of cholesterol, called LDL that yellow, waxy stuff that sticks to artery walls; a lower level of another type of cholesterol called HDL, the "good-guys" that gobble up the LDL and cart it off to the liver; a high level of free-radicals in the blood; and a high level of another type of blood fat – triglycerides. Doctors now think that LDL cholesterol is not so dangerous unless it is converted into a toxic form by free-radicals in your blood. High triglycerides combined with poor LDL/HDL levels quadruples the risk of heart attack. Eating meat sends a two power punch to your system. First, fats from meat and dairy products are the most likely things to send your LDL level soaring. Second, meat (likely non-organic meat) is a source of free-radicals from pesticides, medicines, hormones or other toxins that accumulate in the meat tissues. Rancid meat fat itself can also be a source of free-radicals. Triglycerides come from products that are often consumed with meat in the American culture: refined flour, refined sugar and excessive alcohol. It fact, one school of thought believes that it is not even meat, but the combination of meat fat with refined flour and sugar that is the real culprit in heart disease!

High meat consumption has also been linked to various types of cancers: breast, prostate, colon, pancreas, and non-Hodgkin's lymphoma. According to a six-year study of 14,000 women published in Epidemiology (July 1994), women who ate meat once a day had nearly double the risk of breast cancer of women who said they ate red meat maybe only once a week. Prostate cancer, the second leading cause of male cancer deaths in the U.S., claims the lives of some 35,000 men annually. A four-year Harvard study of 48,000 men found that those who consumed the most red meat – beef, pork, lamb, processed meat, bacon and hot dogs – had the highest risk of ending up with an advanced, or fatal, case of prostate cancer. If your breasts and prostate stay healthy, the biggest cancer risk is to your large bowel or colon. One six-year Harvard study of 90,000 women, found that no amount of red meat was safe – or unable to increase the risk of colon cancer. The study found that those who ate a main dish of red meat daily were 250% more likely to develop colon cancer than those eating meat less than once a month. But even those eating red meat infrequently – once a week or once a month – were still 40% more likely to get colon cancer than those who ate red meat less than once a month. Others studies have found a higher risk of pancreatic cancer (50%) and of non-Hodgkin's lymphoma (73%) with high meat consumption.

Solution: Substitute dried beans and fish for red meat. Oysters, clams, crabs and mussels actually can improve the cholesterol picture, and shrimp will not raise levels! Don't be afraid of eggs, but don't overdo it! In fact you need some high-cholesterol food (shellfish, eggs, liver) in order

⊬ MEAT (Continued)

leading to liver damage. Also, don't overdo the low-fat craze. According to Harvard professor, Frank Sacks, M.D., a very-low-fat diet that restricts calories from fat to 10 percent or less is likely to ruin your good HDL level, leaving you as vulnerable to heart disesase as before. He recommends that 35 to 40 percent of calories should come from fat, but mostly from monounsaturated fats like olive oil. (Fats from corn oil or margarine is not a monounsaturated oil, and is not good. See *Oils.*) Other foods that help your cholesterol picture include oat bran, raw onion, garlic, almonds, avocados, apples and grapefruit. Also, foods high in antioxidants – like Vitamin C, Beta-Carotene, Coenzyme Q-10, Vitamin E and Selenium – also help detoxify bad cholesterol (even reversing clogged arteries) and prevent or thwart the spread of cancer. These include fresh, organic fruits and vegetables like carrots, cabbage, broccoli, collards, spinach, kale, tomatoes, all yellow-orange vegetables, sweet potatoes, beans, cauliflower and fruits, especially citrus. Wheat bran is good to prevent colon cancer. As much as possible, stay away from processed sugars and wheat products. Indigenous cultures that do not yet have these introduced into their diets have remarkably low cancer rates.

⊬ GENETIC-ENGINEERING

[ ] *Do you eat genetically engineered foods?* This is a trick question. You have no way to know whether you are eating them or not. The FDA does not require genetically engineered food to be labeled now. What is currently on the market? Dairy products from cows injected with genetically altered hormone, and tomatoes, corn, squash, potatoes, soybeans and canola. The list is growing fast. Genetic engineering is a radical new technology vastly different from cross-breeding or other genetic manipulations we have done to crops or domestic animals in the past. This technology disregards the reproductive boundaries set in place by nature. It cuts genes from one species and injects them into another, creating life forms that never existed before on the planet. We have begun in a sense to play "God" with the plant and animal kingdom. Many scientists are concerned that manipulating life at this most fundamental level could bring a dangerous chain reaction of consequences in the balance of nature or even in our own disease processes. There is some substance behind their concerns. In 1989, a bacteria genetically engineered to produce a food supplement, tryptophan, also produced a toxic contaminant that killed 37 people and permanently disabled 1,500 others. A genetically engineered bacterium developed to aid in ethanol production, produced residues that made the land infertile. New corn crops planted on this soil grew three inches tall and fell over dead. It was found that a soybean engineered to contain a new protein from the Brazil nut caused the same allergic reaction to the serum of people allergic to the nut. (The allergic people were not actually fed the new soybean, or the reaction could have killed them.) Finally, evidence exists that genes can escape into fertile wild relatives of genetically engineered organisms, creating potentially harmful species.
Solution: Write to your Congressman to demand labeling of genetically engineered foods. For more information contact Mothers for Natural Law (See *Further Resources* in Appendix.)

⊬ SWEETENERS

[ ] *Do you eat much NutraSweet (aspartame) or sugar (sucrose)?* Aspartame is a so-called natural sweetener made of phenylalanine and aspartic acid, which are naturally occurring substances broken down in the body into the same amino acids found in protein food. The aspartame compound is not found in nature, however, and problems are associated with its use. Many people report headaches or "fuzziness," cramps and diarrhea, mood swings and depression or high blood pressure. Aspartame is known to cause birth defects if consumed heavily while a woman is pregnant. People with a genetic disorder called phenylketonuria can experience brain seizures and mental retardation if they consume large doses of phenylalanine. Moreover, an independent board of inquiry that reviewed aspartame in 1980 concluded "...aspartame, at least when given in the huge quantities employed in the studies, may contribute to the development of brain tumors."

White sugar has forty-six calories per tablespoon and provides almost no vitamins or nutrients whatsoever. Conversely, it is a net depleter of nutrients from the body. As with other simple carbohydrates, nutrients must be taken from your blood or even your bones to digest and assimilate sugar. Moreover, for about four hours after you eat sugar, the ability of white blood cells to attack bacteria is inhibited. Overconsumption of sugar is linked to many diseases, including diabetes, heart disease, high blood pressure, osteoporosis, obesity, and vaginal yeast infections, not to mention tooth decay and hyperactivity.

Sugar or sucrose comes in many forms, some worse than others. These include molasses, unprocessed cane and brown sugar, and corn, sorghum, maple or cane syrup. At least some of these less processed forms of sucrose contain a few nutrients, but they still have many of the drawbacks of sugar if consumed in large quantities. They are, however, less contaminated than white sugar, which has usually been sprayed with pesticides and chemically bleached.

Solutions: Stay away from processed foods containing aspartame, sugar or high-fructose corn syrup. Eat naturally sweet foods. Use one or more of the many alternatives to artificial sweeteners and sugar (sucrose) listed.

Barley malt is a natural sweetener that can be found as a syrup or a powder. The powder is about 2000 percent sweeter than sugar, so substitute only two to two and a half teaspoons for a cup of sugar in a recipe. Alternately, the syrup is only forty percent as sweet; substitute two cups of syrup for each cup of sugar, and cut the liquids in the recipe by one fourth cup for each three quarter cup used.

Date sugar is made from dates and can be substituted for sugar one to one. It is especially good in recipes.

Fruit sweetener can be bought as a thick syrup made from unsweetened fruits, or make your own by draining the syrup from frozen fruit. Add an equal amount of clover honey and one quarter teaspoon of vitamin C powder. Use this liquid in a recipe like honey, substituted cup for cup for sugar, but reduce other liquid by one third.

Honey is made of glucose and fructose and is generally the least chemically contaminated. Ayurveda does not recommend the use of honey for cooking, however.

Stevia is a South American herb 80 to 100 times as sweet as sugar, yet contains less than a calorie per serving Lobbying by the artificial sweetener industry has made it illegal to sell now.

⊦ WATER

[ ] *Do you drink or cook with unpurified tap water?* Most Americans take it for granted that the water from their tap is safe to drink. However, frequently it is not. (If your water from your own deep well is not located where contaminants might migrate from leaking gasoline storage tanks, landfills, etc., you could be one of the exceptions.) Your water could comply with all EPA and state regulations, and still be unfit to drink. For example, the EPA allows cities to average levels of a certain class of carcinogens (trihalomethanes/THMs) over a year. Your city could have THMs at levels of thirty percent above the health limit and still be in compliance. In addition, the EPA estimates that at any one time, one in ten of 57,000 community water systems is in violation of standards. (Some of these are administrative violations.) The Natural Resources Defense Council reports some quarter million violations of the federal Safe Drinking Water Act in 1991 and 1992, affecting nearly half the U.S. population. What does this mean to you? The federal Center for Disease Control estimates that nearly a million people a year become ill by drinking contaminated water and about 1,000 of those die. Microscopic bacteria and other parasites in water are responsible for one in three cases of gastro-intestinal illness. One in six Americans drink water with excessive amounts of lead. Lead reduces the attention span and learning ability in children, and contributes to high blood pressure, strokes, heart attacks and premature deaths in men. Industrial pollutants, arsenic, THMs and chlorine, aluminum, pesticides, volatile organic chemicals (VOCs) and radiation have been linked to cancer,

heart disease, impaired mental functioning, Alzheimer's, birth defects or lowered fertility. Currently, it would be exorbitantly expensive for your local water utility company to test for, much less remove, all of these contaminants. Moreover, some pollutants, such as lead, can be picked up in the pipes right in your home. *Solution:* First, find out the most likely contaminants in your area. Talk to local environmental groups and state or local enforcement agencies. Write your public water utility for test results on your water supply. However, remember that these are annual averages. Find out if your house has pipes with lead soldering. Once you know likely contaminants in your water supply, you may want to have your water tested for these particular contaminants. Watch out for water companies who say they will test your water for free. They usually only perform tests for hardness, pH and chlorine, which tells you very little about whether your water is safe to drink. If you are on city water, you can be assured that your water has enough chlorine in it to disinfect a swimming pool, especially in summer. Accredited labs can test your water. Different tests vary from about $20 to $150. (See *Further Resources.*) Once you know your water problems, you can decide the type of filtration system that can best address them. Certain types of problems, such as arsenic, aluminum or mercury, can only be adequately handled by reverse osmosis or distillation. Chlorine, THMs and VOCs require carbon filtration. Bacterial contamination might require ultraviolet. (See *Water and Water Purification* in Appendix.) Ideally, install a whole house system and one for drinking water connected to a special faucet at your kitchen sink. (See *Product Sources,* under *Food.*)

35

✛ PET FOOD

[ ] *Do you feed your pet commercial dry or canned pet food?* The quality of commercial brands can vary, but many can be contaminated with the same things that contaminate processed human foods. (See *Food & Pesticides & Synthetic Chemicals*.) These include artificial colors, flavors and preservatives, nitrates, drugs and pesticides. Some companies get their protein from cheap meats (animals that are already dying or diseased), then add sugar, a few nutrients and artificial color so that it looks fresh. To be labeled "nutritionally complete," a pet food only has to contain a dozen or so recommended vitamins and minerals; but there are forty known essential nutrients, not to mention the scores of associated food factors and other trace minerals still under investigation. One author reports that cats lose their immune system after three generations on a strictly canned food diet. Another physician performed a ten-year study of cats fed raw versus cooked meat. The cooked-meat group had health problems, like heart disease and arthritis, that worsened with each generation. The raw-meat group, in contrast, were more resistant to infections, fleas and worms, and were friendlier, even-tempered and well-coordinated. They had few miscarriages and lived much longer. *Solutions: Keep your Pet Healthy the Natural Way* advises raw, organically grown meat for one-half of your dog's diet and three-fourths of your cat's diet. Others advise that even organic meat (if raw) these days brings a high risk of cancer due to the presence of parasites. (See reference to Dr. Clark in *Cancer and Modern Living* in Appendix.) My holistic vet says that occasional raw meat is good for your pet if its immune system is strong. Your pet's diet can and should be supplemented, however, with fresh, organic fruits, vegetables and grains, as well as occasional table scraps. Vegetables should be cut up and eaten raw or slightly cooked. Feed fruits separately. My Rottweiler loves fresh cut broccoli stems, squash, tomatoes, potatoes and cauliflower. A friend's cat's favorite is steamed fish sprinkled with kelp. Her dog loves avocado. Some California purebred dog breeders feed a first meal of barley flakes or oats soaked in raw milk for four hours (or you can cook), mixed with one tablespoon of cold pressed oil, some cornmeal, kelp and sprouts or fresh green herbs. My dog loves organic oats cooked with organic milk. Supplement your pets' diet with raw marrow bones (no rawhide or plastic please) and always feed your pets purified water. If you feed your pet only dry food, you need to supplement the diet with oils. I have been told to supplement the feed of my large dog with one tablespoon of flax seed or fish oil a few times a week. (See *Oils* in Chapter VI. Food.) If your pet's diet is mainly dry food, give pet vitamins from whole plant concentrates grown on organic farms or powdered blue green algae. These can supply trace elements, vitamins and amino acids (algae). If you buy dry food, buy a natural one without artificial flavorings – the fresher the better (PHD and Flint River Ranch). Always make any changes in your pet's diet gradually by mixing the old in increasing proportions with the new. My holistic vet recommends alternating a lamb with a chicken formula. Always consult your local holistic veterinarian about diet. (See *Food –Pet Food* and *Consumables – Nutritional Supplements* under *Product Sources* in Appendix.)

APPENDICES

Product Sources

House Structure

Air Krete & Natural Cork Insulation
Air-Krete, East Brutus,
Weedsport, NY 13166
(315) 834-6609

Rector Mineral Trading Corp.
9 W. Prospect Ave.
Mt. Vernon, NY 10550
(914) 699-5755

Air Purification
(See *Air Purification* in Appendix)
Green Globe Products, Inc.
P.O. Box 723
Silver Spring, MD 20918
(301) 681-3489

Sealers (Safe Seal), & Tub Spout
Green Globe Products (See above)

Healthy Home Center
1403-A Cleveland St.
Clearwater, FL 34615
(813) 447-4454

Wood Products
Agri-Boards
1500 S. Main St.
Fairfield, IA 52556
(515) 472-0363

Eagle Cabinents & Construction
HC62 Box 29
Flippin, AR 72634
(870) 453-3245

Envt'l Construction Outfitters
44 Crosby St.
New York, NY 10012
(800) 238-5008 or (212) 334-9659

Medite Corporation/Sales
P.O. Box 4040
Medford, OR 97501
(800) 676-3339 or (541) 773-2522

Wood Stoves & Fireplaces
Pyro Industries, Inc. (*Pellet Stoves*)
695 Pease Rd.
Burlington, WA 98233
(360) 757-9739

Abundant Health (*Smokeless Stove*)
P.O. Box 534
La Mesa, CA 91944
(619) 462-6468

Ruegg Fireplaces
976 Rt. 22
Bridgewater, NJ 08807
(201) 526-0309

Vermont Castings (*Fireplace Inserts*)
Rt. 107
Bethel, VT 05032
(802) 234-2300

Waterford Irish Stoves
16 Airpark Rd.
Lebanon, NJ 03784
(603) 298-5030

Home Interiors

Carpet and Rugs (Natural Fiber)
Carousel Carpet Mills
1 Carousel Ln.
Ukiah, CA 95482
(707) 485-0333

Garnet Hill
Box 262 Main St.
Franconia, NH 03580-0262
(800) 622-6216

Natural Home
P.O. Box 1677
Sebastopol, CA 95473
(707) 824-0914

Sinan Company (*Wool*)
P.O. Box 857
Davis, CA 95617-0857
(916) 753-3104

Carpet Adhesives and Guard
Child Safe Electric Outlets
Green Globe Products, Inc.
P.O. Box 723
Silver Spring, MD 20918
(301) 681-3489

Healthy Home Center
(813) 447-4454 (See pg. 38)

Floor Coverings/Linoleum
Bangor Cork Corp., Inc.
P.O. Box 125
Pen Argyl, PA 18072
(610) 863-9041

Forbo Industries
P.O. Box 667
Hazelton, PA 18201
(800) 733-3297

Natural Home
(707) 824-0914 (See pg. 38)

EMF Products, Full-Spectrum Light
Whole House Water Purification
Green Globe Products (See above)

General
Envt'l Construction Outfitters
(800) 238-5008 or (212) 334-9659
(See pg. 38)

Paint or Wood Finishers
Green Globe Products (See above)

Healthy Home Center (See above)

Miller Paint Co.
317 SE Grand Ave.
Portland, OR 97214
(503) 233-4491

Natural Choice
1365 Rufina Circle
Sante Fe, NM 87505
(505) 438-3448

Sinan Company
(916) 753-3104 (See pg. 38)

The Old Fashioned Milk Paint Co.
P.O. Box 222
Groton, MA 01450
(508) 448-6336

Natural Wall Paneling & Ceiling Tile
Tectum Inc.
P.O.Box 3002
Newark, OH 43058-3002
(800) 410-1121

Window Glass (For Full-spectrum)
Schott America Glass
3 Odell Plaza
Yonkers, NY 10701
(914) 968-1400 or 969-6100

Home Furnishings

Furniture/Cotton Futons, Mattresses
Allergy Relief Shop
3371 Whittle Springs Rd.
Knoxville, TN 37917
(800) 626-2810 or (423) 522-2795

Allergy Resources, Inc.
P.O. Box 444
Guffey, CO 80820
(800) 873-3529 or (719) 689-2969

High Cotton Company
39 Broadway
Asheville, NC 28001
(704) 253-1138

Janice Corporation
198 Route 46
Budd Lake, NJ 07828
(800) 526-4237

Charles R. Bailey Cabinent Makers
HC62 Box 29
Flippin, AR 72634
(870) 453-3245

Shaker Workshops
P.O. Box 1028
Concord, MA 01742
(617) 646-8985

Garnet Hill
Box 262 Main St.
Franconia, NH 03580
(800) 622-6216

Seventh Generation
1 Mill St., Box A26
Burlington, VT 05401-1545
(800) 456-1177

Feather Beds
Cuddledown
P.O. Box 1910
Portland, ME 04104
(207) 761-1855

Feathered Friends
1415 Tenth Ave.
Seattle, WA 98122
(206) 328-0887

Upholstery
Homespun Fabrics & Draperies
P.O. Box 3223
Ventura, CA 93006
(805) 642-8111

Cotton Bedding/Linens/Towels
Allergy Relief Shop
(800) 626-2810 (See pg. 39)

Garnet Hill (See above)

Karen's Nontoxic Products
322 St. Johns St.
Havre de Grace, MD 21078
(800) 527-3674 or (410) 378-4936

Seventh Generation (See above)

Vermont Country Store
P.O. Box 3000
Manchester Center, VT 05255-3000
(802) 362-4647

Toys
Baby Bunz and Company
P.O. Box 113
Lynden, WA 98264
(360) 354-1320

Toys (Cont.)
Hearthsong
P.O. Box 1773
Peoria, IL 61656-1773
(800) 325-2502

Karen's Nontoxic Products
(See previous column)

World Wide Games
P.O. Box 517
Colchester, CT 06415-0517
(800) 243-9232

Art Paints/Glues
Karen's Nontoxic Products
(See previous column)

Natural Choice
(505) 438-3448 (See pg. 39)

Appliances

Vacuums
Kirby (*Micro-filtration*)
1920 W. 114th St.
Cleveland, OH 44102
(216) 228-2400

Nilfisk of America, Inc. (*HEPA*)
300 Technology Dr.
Malvern, PA 19355
(800) NILFISK or 645-3475

Scovill (*Whole-house*)
NuTone Division
Madison & Red Banks Rds.
Cincinnati, OH 45227
(513) 527-5100

Sinan Company (*HEPA*)
(916) 753-3104 (See pg. 38)

Ultrasonic Humidifiers
Ellis & Watts, Inc./Sales
4400 Glen Willow Lake Ln.
Badavia, OH 45103
(513) 752-9000

Consumables

Cleaners
Brands to look for:
Safe Choice ⊛
Awalan ⊛
Dr. Bronner's
EarthRite ⊛
Earth Friendly Products TM
Ecover ⊛
Granny's
Naturally Yours
Seventh Generation ⊛
Sodasan

Sources:
Green Globe Products, Inc.
P.O. Box 723
Silver Spring, MD 20918
(301) 681-3489

(Awalan ⊛, Granny's)
Allergy Resources, Inc. (See pg. 39)
(800) 873-3529 or (719) 689-2969

(Granny's)
Healthy Home Center
(813) 447-4454 (See pg. 38)

Karen's Nontoxic Products
(800) 527-3674 (See pg. 40)

(Sodasan)
Natural Choice
(505) 438-3448 (See pg. 39)

Naturally Yours Ecolo-clean
1926 S. Glenstone Ave., Box 406
Springfield, MO 65804
(417) 889-3995

Seventh Generation ⊛
(800) 456-1177 (See pg. 40)

(Awalan ⊛)
Sinan Company
(916) 753-3104 (See pg. 38)

Nutritional Supplements and Herbs
Whole, Organic, Effective Brands:
Maharishi Ayur-VedTM Herbs,
Alive EnergyTM, Others (pls. call)
Source: Green Globe Products, Inc.
(See previous column)

Personal Care Products
Brands to look for:
Alexandra Avery
Aubrey Organics ⊛
Aura Glow
Dr. Hauschka
Granny's
Helena Meyer
Logona
Paul Penders
Rachel Perry ⊛
Satin Touch Head and Body Shampoo
Simple Wisdom
The Natural Dentist TM
Thursday Plantation
Walter Rau ⊛
Weleda
Vicco

Sources:
(Aura Glow, Helena Meyer, Satin
Touch, Weleda)
Green Globe Products
(See previous column)

(Ida Grae, Dr. Hauschka, Granny's)
Allergy Resources, Inc.
(800) 873-3529 or (719) 689-2969

(Alexandra Avery, Vicco, Aubrey
Organics ⊛)
Karen's Nontoxics (800) 527-3674

(Dr. Hauschka, Walter Rau ⊛,
Logona)
Meadowbrook Herb Garden
Route 138
Wyoming, RI 02898
(401) 539-7603

Paper Products

Brands to look for:
Envision ⊙
Fort Howard ⊙
Green Forest TM
Seventh Generation ⊙

Sources:
Karen's Nontoxic Products
(800) 527-3674 (See pg. 40)

Green Globe Products, Inc.
(301) 681-3489 (See pg. 38)

Seventh Generation
(800) 456-1177 (See pg. 40)

Car Care

Propylene Glycol Antifreeze Brands:
Sierra ⊙
Sta-Clean ⊙

Safer Car Wax Brands:
Kit Wax ⊙
Turtle Wax ⊙

Naturally Yours Degreaser
Source: Naturally Yours Ecolo-clean
(417) 889-3995 (See pg. 41)

Green Globe Products, Inc.
(301) 681-3489 (See pg. 38)

Pest Control

Pyrethrum Powder/
Diatomaceous Earth:
Allergy Resources, Inc. (See pg. 39)
(800) 873-3529 or (719) 689-2969

Insect Traps:
Gardens Alive!
5100 Schenley Pl.
Lawrenceburg, IN 47025
(812) 537-8650

Insect Repellent:
Brands To Look For:
Green Ban ⊙
Helena Meyer

Sources:
Green Globe Products, Inc.
(301) 681-3489 (See pg. 38)

The Ecology Box
P.O. Box 277
Clinton, MI 49236
(800) 735-1371

Lawn and Garden

Brands to look for:
Agrevo TM
Concern ⊙ by Necessary Organics
Garden's Alive! ⊙
Safer ⊙ Soaps and Attack Products
Surefire TM

Natural Enemies:
Beneficial Insectary
14751 Oak Run Rd.
Oak Run, CA 96069
(916) 472-3715

Natural Fertilizers and Pest Control:
Gardens Alive!
(812) 537-8650 (See prior column)

Large home & garden supply chains now carry the Safer line of Products. Check your local hardware store, nursery or garden center for others.

Pet Care

Non-Toxic Shampoos and Pest Control, Non-toxic Cat Litter:

Karen's Nontoxics
(800) 527-3674 (See pg. 40)

The Natural Pet Care Company
8050 Lake City Way, NE
Seattle, WA 98115
(800) 962-8266

Food

Look for foods that are certified organic and/or that have simple, easy to understand ingredients, i.e. have no dyes, preservatives or other additives. It's best to eat mostly fresh, whole organic foods like fruits, vegetables and nuts and juices made fresh with a juicer. A small selection of processed foods follow:

Cereals/Snacks: Barbara's ⊛
Health Valley ⊛
Arrowhead Mills ⊛
Life Stream TM
Garden of Eatin' ⊛
Bearitos ⊛
Dairy: Kate's (Butter)
Stoneyfield Farm ⊛ (Yogurt)
Horizon ⊛ (Milk, Eggs, Butter)
Juniper Valley (Milk)
Natural by Nature TM (Milk)
The Happy Hen (Eggs)
Meats: Coleman Natural Beef ⊛
Bell and Evans ⊛ (Chicken)
Smithfield Lean Generation TM (Pork)
Desserts: Sweet Nothings ⊛
Westbrae Natural ⊛
Hain Pure Foods ⊛
Meals: Amy's Kitchen
Cascadian Farm ⊛
Oils: Spectrum Naturals ⊛
Hain Pure Foods ⊛
Pasta: DeBole's ⊛
Michelle's Natural
Millina's Finest TM
Sauces: Enricos
Garden Valley Naturals TM
Frozen Vegetables:
Cascadian Farms ⊛

Allergy Resources, Inc.
(800) 873-3529 or (719) 689-2969

Mountain Ark Trading Company
799 Old Leicester Hwy.
Asheville, NC 28806
(800) 643-8909

Water Purification Systems
(See *Water Purification* in Appendix)

Source:
Green Globe Products, Inc.
P.O. Box 723
Silver Spring, MD 20918
(301) 681-3489

Pet Food

Brands to look for:
*Flint River Ranch
INNOVA
Lick Your Chops
Natural Life Pet Products ⊛
Nature's Recipe TM
Nutro ⊛ Natural Max
Precise ⊛
PetGuard ⊛
*PHD "Perfect Health Diet"
Solid Gold ⊛
Wysong ⊛ (Amish raised chicken)
Wow-Bow

Sources:
Flint River Ranch
1243 Columbia Ave., St. B-6
Riverside, CA 92507
(909) 682-5048

Halo Pet Products
3438 E. Lake Rd., #14
Palm Harbor, FL 34685
(800) 842-6624

The Natural Pet Care Company
(800) 962-8266 (See pg. 42)

PHD "Perfect Health Diet"
P.O. Box 8313
White Plains, NY 10602
(For orders) (800) 743-1502
(For information) (800) 320-7062

Wow-Bow Distributors
13-B Lucon Dr.
Deer Park, NY 11729
(800) 326-0230 or (516) 254-6064

Further Resources
ORGANIZATIONS

Asbestos

Asbestos List (EPA-certified asbestos detection agencies)
Consumer Product Safety Commission
Washington, DC 20207
(301) 504-0580 (Public Affairs)

Building Construction (Healthy)

Center for Resourceful Building Technology
P.O. Box 100
Missoula, MT 59806
(406) 549-7678

Center for Sustainable Building
221 Concord Ave.
Cambridge, MA 02138
(617) 868-7788

SEAS/Sustainable Environment Association
Paul Bierman-Lytle (designer)
(203) 966-3541

What's Working
David Johnston
(303) 444-7044

Consumer Choices

The Ecology Center
2530 San Pablo Avenue
Berkeley, CA 94702
(510) 548-2220

EcoStewards Alliance
5765-F Burke Centre Pkwy., St. 321
Burke, VA 22015-2233
(703) 503-8232

Global Action Plan
P.O. Box 428
Woodstock, NY 12498
(914) 679-4830

Mothers for Natural Law
P.O. Box 1177
Fairfield, IA 52556
(515) 472-2809

Carpet Cleaning

Institute of Inspection Cleaning and Restoration Certification (IICRC)
(800) 835-4624

Consciousness/Human Potential

Brian Tracy International
462 Stevens Ave., St. 202
Solana Beach, CA 92075
(619) 481-2977

Income Builder's International (IBI)
(205) 837-1130

Transcendental Meditation Program
(888) LEARNTM or 532-7686

Dentistry (Holistic)

Clifford Consulting and Research
(719) 550-0008

General

Environmental Defense Fund
1875 Connecticut Ave., St. 1016
Washington, DC 20009
(202) 387-3500

Natural Resources Defense Council
40 W. 20th St.
New York, NY 10011
(212) 727-2700

Worldwatch Institute
1776 Massachusetts Ave., NW
Washington, DC 20036
(202) 452-1999

Edu-K/Muscle Testing

Edu-K Foundation
(800) 356-2109

Hazardous Wastes

Citizen's Clearinghouse for
Hazardous Wastes
P.O. Box 6806
Falls Church, VA 22040
(703) 237- 2249

Environmental Hazards Management
Institute
P.O. Box 932
Durham, NH 03824
(603) 868-1496

EPA Hazardous Waste Hotline
(800) 424-9346

Household Hazardous Waste Project
1031 Battlefield, St. 214
Springfield, MO 65807
(417) 889-5000

The Local Government Commission
1414 K St., St. 250
Sacramento, CA 95814
(916) 448-1198

Mass. League of Women Voters
133 Portland St.
Boston, MA 02114
(617) 523-2999

Health and Environment

American Academy of Environmental
Medicine
4510 W. 89th St.
Prarie Village, KS 66207
(913) 642-6062

Center for Mind/Body Medicine
5225 Connecticut Ave., NW, St. 414
Washington, D.C. 20015
(202) 966-7338

Environmental Health Center
8345 Walnut Hill Ln., St. 205
Dallas, TX 75231-4262
(214) 368-4132, ext. 146

Gary Null and Associates
P.O. Box 918 Planetarium Station
New York, NY 10024
(212) 799-1246

Maharishi Ayurvedic Medical Center
679 Georgehill Rd.
Lancaster, MA 01523
(508) 365-4549

Organic Gardening

Rodale Institute
222 Main St.
Emmaus, PA 18098-0099
(610) 683-6383

Community Alliance with Family
Farmers
P.O. Box 363
Davis, CA 95617
(916) 756-8518

Pesticides

Bio-Integral Resource Center
P.O. Box 7414
Berkeley, CA 94707
(415) 524-2567

EPA Pesticide Hotline
(800) 858-7378

International Alliance for Sustainable
Agriculture
1710 University Ave., SE, Rm. 202
Minneapolis, MN 55414
(612) 331-1099

Mothers and Others
40 W. 20th St.
New York, NY 10011
(212) 242-0010

National Coalition Against the Misuse
of Pesticides
530 Seventh St., SE
Washington, DC 20003

Pesticides (cont.)

NW Coalition for Alternatives to
Pesticides (NCAP)
P.O. Box 1393
Eugene, OR 97440
(503) 344-5044

Pesticide Action Network
116 New Montgomery St., St. 810
San Francisco, CA 94105
(415) 541-9140

Radon

National Safety Council
P.O. Box 33435
Washington, DC 20077
(800) SOS-RADON

Technological Advances

Green Globe Products
P.O. Box 723
Silver Spring, MD 20918
(301) 681-3489

Testing

CasChem Laboratories, Inc.
1712 Ira Turpin Way
Canton, OH 44705
(800) 800- 6052 or (330) 588-8378
(*Tests soils, paint, asbestos, etc.*)

Suburban Water Testing Labs
4600 Kutztown Rd.
Temple, PA 19560
(800) 433-6595

Watercheck National Testing Labs
6151 Willson Mills Rd.
Cleveland, OH 44133
(800) 426-8378

For discounts, call (301) 681-3489
Green Globe Products (See above)

Healthy Spaces (*Home Audits*)
P.O. Box 69
Lovettsville, VA 20180
(800) 290-4436 or (540) 822-4010

Self-Tests/Monitoring
The following items are available at
your local hardware stores:
o Carbon monoxide detectors that
resemble smoke detectors ~$30-$60
o Inexpensive lead-testing materials
for testing walls, etc ~$8
o Radon testing kits ~$25 (Radon
testing kits are also available from the
National Safety Council for $9.95.)

Water Issues

Clean Water Action Project
4455 Connecticut Ave., NW, St. 8300
Washington, DC 20008
(202) 895-0420

EPA Safe Drinking Water Hotline
(800) 426-4791

Water Environment Federation
601 Wythe Street
Alexandria, VA 22314-1994
(703) 684-2400

Water Quality Association
4151 Napier Rd.
Lisle, IL 60532
(708) 505-0160

World Peace

Carter Center
1 Copenhill Ave.
Atlanta, GA 30307
(404) 420-5100

World Peace Institute or
The Oneness Institute
6339 East Greenway Rd, St. 102-346
Scottsdale, AZ 85254
(602) 905-7170

BOOKS AND TAPES
Building for Health and Ecology

Sourcebook for Sustainable Design
Boston Society of Architects
(617) 951-1433

Sustainable Building Sourcebook
City of Austin (512) 499-3500

Cancer

Books
Clark, H.R., 1993. *The Cure for All Cancers.* San Diego, CA: ProMotion Publishing, 511 p. (800) 231-1776

Eiden, W.K. 1996. *The Man Who Cures Cancer.* Bethesda, MD: Be Well Books, 352 p. (800) 879-4214

Thomas, R., 1993 *The Essiac Report.* LA, CA: Alternative Treatment Info. Network, 109 p.+ (310) 278-6611

Tapes
Alternative Approaches to Disease Cancer: The Natural Approach
Videos by Gary Null (212) 799-1246

Consumer Choices

Anderson, B.N., 1990. *Ecologue. The Environmental Catalogue and Consumer's Guide for a Safe Earth.* New York: Prentice Hall, 255 p.

Food and Health

Carper, J., 1993. *Food - Your Miracle Medicine.* New York: HarperCollins Publishers, 528 p.

Cousens, G., 1995. *Conscious Eating.* Patagonia, AZ: Essene Vision Books, 527 p. (800) 754-2440

Null, G. 1995. *Nutrition and the Mind.* New York: Four Walls Eight Windows, 304 p.

Steinman, D., 1992. *Diet for a Poisoned Planet.* New York: Ballentine Books, 326 p.

Electromagnetic Fields

Brodeur, P., 1993. *The Great Power-Line Cover-up: How the Utilities and the Government are Trying to Hide the Cancer Hazard Posed by EMFs.* Boston: Little, Brown, and Co., 326 p.

EMFs in Your Environment: Magnetic Field Measurements of Everyday Electrical Devices. Washington, D.C.: GPO, 21 p.

Environment

Gore, A., 1992. *Earth in the Balance.* Boston, New York, London: Haughton Mifflin Co., 407 p.

Hawkin, P., 1993. *The Ecology of Commerce.* New York: Harper Business, 250 p.

Health and Longevity

Chopra, Deepak, 1991. *Perfect Health.* New York: Harmony Books, 327 p.

Gordon, James, S., 1996. *Manifesto for a New Medicine.* USA: Addison-Wesley Publishing Co., Inc., 358 p.

Null, G., 1994. *The 90s Healthy Body Book.* Deerfield Beach, FL: Health Communications, Inc., 256 p.

Thomas, J., 1995. *Young AGAIN!* Kelso, WA: Plexus Press, 384 p.

Home Interiors

Linn, D., 1995. *Sacred Space.* New York: Ballantine Books, 308 p.

Indigenous Views

Eirira, A., 1992. *The Elder Brothers.* New York: Alfred A. Knopf, 243 p.

Maels, Thomas E., 1991. *Fools Crow Wisdom and Power.* Tulsa, OK: Council Oak Books, 203 p.

Morgan, M., *Mutant Message/Down Under.* New York: HarperCollins Publishers, Inc., 187 p.

Negative Ions and Health

Soyka, F. with Edmonds, A., 1977. *The Ion Effect.* New York: Bantam Books, Inc. (No longer in print).

Organic/EnlightenedFarming/Lawn

Borman, F.H. and G.T. Geballe, 1995. *Redesigning the American Lawn: A Search for Environmental Harmony.* New Haven, CT: Yale University, 166 p.

National Organic Directory. Yearly. Davis, CA: Community Alliance with Family Farmers, ~ 400 p.

Fukuoka, M., 1985. *The Natural Way of Farming: the theory and practice of green philosophy.* Tokyo: Japan Publishers, 280 p.

Hawken, P., 1976. *The Magic of Findhorn.* New York: Bantham Books, 343 p.

Personal and Spiritual Growth

A Course in Miracles. 1985. Glen Ellen, CA: Foundation for Inner Peace, 478 p.

Burns, David D., 1980. *Feeling Good. The New Mood Therapy.* New York: Penguin, 413 p.

A. Wally Minto, 1994. *Communication & Understanding in Relationships.* Dallas, 139 p. (916) 279-2226

Maharishi Mahesh Yogi, 1983. *Maharishi Mahesh Yogi on the Bhagavad-Gita.* East Rutherford, N.J.: Penguin Books, 494 p.

Peck, M. S., 1978. *The Road Less Traveled: A New Psychology of Love, Traditional Values and Spiritual Growth.* New York: Simon and Schuster, 316 p.

Yogananda, P., 1985. *Autobiography of a Yogi.* Los Angeles, CA: Self-Realization Fellowship, 591 p.

Tapes
Oneness: A Dynamic Consciousness for a New Century. (602) 905-7170
Sidhana: The Path to Enlightenment. Penny Price Video (516) 897-3120
Videos by Brian Tracy
The Psychology of Achievement
The Universal Laws of Success and Achievement (800) 945-3132
The Deepak Chopra Video Collection
The Seven Spiritual Laws of Success
The Way of the Wizard
Overcoming the Fear of Death
Explorations into Consciousness
Mystic Fire Video (800) 292-9001

MAGAZINES

The EarthWise Consumer
P.O. Box 1506
Mill Valley, CA 94942

Environ, A Magazine for Ecologic Living
P.O. Box 2204
Fort Collins, CO 80522

Environmental Building News
RR1 Box 161
Brattleboro, VT 05301

Muscle Testing/Educational Kinesiology

Muscle testing is a technique for learning information that circumvents the thinking mind. It's like a simple lie detector test, or a way of gaining information from all the levels of intelligence connected to the body, mind and spirit. Muscle testing is a way of getting yes/no or true/false answers by testing the strength or weakness of muscle groups in response to a question or a statement. For example, if your name is Carol but you say, "My name is Sam," your muscles will respond with weakness to a challenge. No matter how hard you try, you will not be able to lock the muscles as firmly as you could if you said, "My name is Carol."

You can use muscle testing to see if a household or herbal product is right for you. The products can be held in the hand or against the solar plexus, and a partner can check the muscles to see if they lock. If they lock, the product is beneficial. The muscles will not lock if the product is harmful or not particularly beneficial. If you are suffering from an environmental illness, you can also ask if the problem is environmental in nature and get a yes/no response from the body. You can ask almost any kind of question about a problem or a product. It is best to work with an experienced practioner in the beginning, but some brief instructions follow.

Although any muscle can be used in the test, the muscle normally tested is the deltoid muscle of the arm. To conduct a muscle test, hold a product against your solar plexus and extend your arm in front of you and slightly to the side at a 90° angle. Rather than holding a product, your partner or you can ask a question. Have a partner apply pressure with two fingers about four inches behind your wrist. The pressure should be steady and firm, not a sudden force. Now do your best to hold your arm straight out against the pressure that your partner is applying. If you can lock your arm, your body has indicated a preference for the product, or an affirmative to the question. If the answer is negative, you will not be able to hold up the arm. Really. Try it a few times with test questions that you know to be true or false and see what I mean. Self-muscle tests can be conducted by trying to pull your thumb apart from your third finger on one hand, with the thumb and third finger of the second hand. Again, holding means yes, and pulling apart means no.

Muscle testing can give inaccurate answers if both persons don't "have their clearing." This entails having sufficient movement, water and stimulation of the "brain buttons." To stimulate the brain buttons, hold the left hand over your navel, and rub deeply with the thumb and third finger of the right hand just below the collarbone, to the right and left of the sternum.

Muscle testing is used in Educational Kinesiology (Edu-K), a system of accelerated learning that removes learning blocks. Edu-K uses the whole brain through movement repatterning with movements and techniques taken from yoga and acupuncture. It was developed by Paul and Gail Dennison and has experienced phenomenol growth in use by educators across the nation. For more information, see *Further Resources.*

Instructions adapted from a guide by Deborah Smith that can be ordered at http://svr.com/starsmith/. She may be contacted at Email: starsmth@oneimage.com.

Recipes

Commonly Used Ingredients

1. Heinz White Vinegar
2. Baking Soda
3. Washing Soda
4. Borax*
5. Lemon Juice
6. Salt
7. Liquid Soap

Occasional Ingredients

1. Ketchup
2. Worcestershire Sauce
3. Mayonnaise
4. Olive Oil
5. Trisodium Phosphate*
6. Toothpaste

Sources: Golden Empire Health Planning Center, 1990.
Dix, Karen, 1990. (See references).

Natural Cleaning Implements

1. Instead of Paper Towels -
   Cellulose Sponge Cloth
   100% Cotton Towels
2. Instead of Synthetic Sponges -
   Cellulose Compressed Sponges
   French Cotton Sponges
3. Polishing Clothes -
   Cheesecloth
   Rags, Old Cotton Clothes
4. Vegetable Brush -
   Natural Bristle Brush
5. Pumice Stone -
   Removes stains from procelain

* Products marked with an asterisk are not considered nontoxic. They are single product ingredients, however, derived from natural sources. You can easily avoid them in the future if you have an adverse reaction.

These recipes are especially important for persons with chemical sensitivities due to an overexposure to toxic chemicals. For further information: See Berthold-Bond, A. 1990. *Clean and Green.* Woodstock, NY: Ceres Press, 162 p. (See *References.*)

Vinegar Solution 1: 1 part water/1 part vinegar mix (1 gal.Heinz white with 2 tbsp. Eden or other organic Apple Cider Vinegar with the "Mother" pulp at the bottom).
Vinegar Solution 2: 1 part water/1 part vinegar mix (1 gal.Heinz white/4 oz. Eden).
Baking Soda Solution: 8 ounces baking soda in 1 gallon warm water.

Glass Cleaners
o 1/4 to 1/2 cup white vinegar in a quart of warm water. Wipe with a cotton cheesecloth.

Painted Wood
o 1 tsp. of washing soda in a gallon of hot water

Drains
o Pour boiling water down the drain once a week.
o If clogged, pour 1/4 cup of baking soda, followed by 1/2 cup white vinegar. Let set, covered, a few minutes. Follow with a kettle of boiling water. You are creating an intense bubbling; some authorities are concerned about damage to pipes.
o If the above does not work, use a professional snake.

Oven Cleaners
o Sprinkle pure water followed by layers of baking soda. Rub gently with very fine (ooooo) steel wool pads.
o Mix 3 tbsp. of washing soda with 1 quart of warm water.

Toilet Bowl Cleaners
o Sprinkle baking soda, scrub with toilet brush.
o For disinfecting, use 1/2 cup borax* to 1 gallon of hot water.

Scouring Powder
o Use straight borax*, baking soda or salt on a wet sponge.
   For whitening, use lemon juice.
o A pumice stone will remove stains from procelain toilets, tubs or sinks.

Linoleum Floors
o Mop linoleum floors with 1 cup white vinegar mixed with 2 gallons of water. Polish with club soda.

Wood Furniture and Floor Polish
o Use mayonnaise.
o 3 parts olive oil with with 1 part Heinz white vinegar.
o 2 parts olive oil with 1 part lemon juice.

Mold or Mildew
o Dampen sponge and wipe borax* on mold prone areas.
o Mix 1 part white vinegar and 1 part water. Spray on mold, let set and wipe.

Germ Disinfectant
o 1/2 cup borax* in 1 gallon of hot water. (Hospital tested and approved.)

Air Fresheners/Deodorizers
o Sprinkle borax* or baking soda in odor prone areas.
o Leave an opened box of baking soda in room or refrigerator.
o Vinegar in soapy water removes onion odors.
o Put cloves or cinnamon sticks, or slices of lemon/grapefruit/ or orange, in enough
   water to simmer an hour or so.
o Put out fresh or dried flowers; grow plants.

Fireplace, Concrete, or Brick Cleaner
o Trisodium phosphate (TSP)*

Polishes: Brass   o Rub with Worcestershire sauce or ketchup, let sit, polish dry
          Copper o Rub with vinegar and salt
          Silver   o Place cut up aluminum foil strips in salt water (1 tbsp. salt/ 1
pint of water). Place the article in the solution for 5 minutes, remove and polish.

All Purpose Spot Remover
o Dissolve 1/4 cup of borax* in 2 cups of cold water. Sponge it on and let it sit
until it dries, or soak the fabric in the solution before washing it in cold water. This
works well for blood, chocolate, coffee, mildew, mud and urine.

Energy and Nutritional Supplements

Though nutrition may appear to be slightly off the topic of home environmental factors affecting health, herbs and nutritional supplements are important for detoxification, energy level maintenance and health in general. These products will be particularly important as long as modern agriculture raises foods on depleted and unhealthy soils. Optimal health requires intake of all the necessary vitamins, proteins, enzymes, trace minerals and associated food factors frequently missing from food which has been heavily processed or grown on depleted soils created by non-organic farming. For example, because of the widespread use of artificial fertilizers containing potassium, potassium has generally replaced magnesium in our soils and in our food. Magnesium is very important to optimal health. Persons with insufficient magnesium find it difficult to relax and handle the everyday stress of life. Foods harvested from low on the food chain in large bodies of water, such as blue green algae, as well as concentrated and organically grown herbs and plants, make excellent nutritional supplements. Certain herbs in particular have an ability to excite the body's energy-creating metabolic pathways directly. Here are the best.

(1) **Blue green algae, spirulina, chlorella.** Often referred to as "super-foods," they contain in a compact package the eight essential amino acids and the vitamins, minerals and enzymes needed for optimal health. Chlorophyll also helps the body eliminate toxins and their detrimental effects. Wild-grown is best, like the blue green algae from Klamath Lake in Oregon.

(2) **Flaxseed oil and black current seed oil.** These oils supply essential fatty acids (EFAs). Despite the low-fat fads, a certain amount and type of fats are absolutely essential for health. EFAs in particular are consumed rapidly by the body's energy cycles. Flaxseed oil supplies Omega-3 fatty acids which increase metabolic rate. Recommended usage: 2 tbsp. per day on food.

(3) **Asian, American and Siberian ginseng and licorice root.** These herbs can dramatically affect energy levels through prevention of adrenal depletion. The adrenal glands help control stress, and repeated stress depletes them. The three types of ginseng have different uses, so consult a good book on herbs or a professional herbalist. Use only whole licorice root, either stick or powdered.

(4) **Gingko biloba.** Few of us can feel energized when our minds are sluggish. This herb increases circulation to the brain, is good for Alzheimer's and detoxifies blood vessels.

(5) **Ginger and cayenne.** These spices improve cardiovascular health through increasing blood flow and cholesterol metabolism, and by lowering LDLs. They also aid in digestion and help prevent gastrointestinal and genito-urinary infections.

If you suffer from chronic fatigue and your doctor cannot find a medical reason, in addition to using these herbs, stay away from refined white sugar and flour, including pasta; and caffeine, including chocolate. You may also have food allergies or toxic overload. (See *Product Sources, Consumables – Nutritional Products* in Appendix.)

Liver and Gallbladder Detoxification

This flush is designed to detoxify and restore the normal functional capacity of these organs. Use fresh-squeezed juice made with organic or home-grown vegetables. Consult with your doctor before trying this regimen.

LIVER PURGE
Do the following for one month:
  Garlic. Take 2 capsules morning and evening (See *Product Sources, Consumables*)
  Punch. Squeeze 6 lemons, 12 oranges, 6 grapefruits; add water to make 1 gallon of liquid. Drink at least 3 glasses per day.
  Fast. On the 7th day of each week, fast on 1 quart of carrot juice, 1 pint of celery juice, 1/2 pint of beet juice. Mix them together; drink slowly throughout the day.

GALLBLADDER FLUSH
The following steps should take about a week and follow the liver purge:
  Step 1. Monday through Saturday noon, drink as much apple juice or apple cider as your appetite will permit in addition to regular meals and prescribed supplements. The apple juice should be organic, coarse, unfiltered and free of additives and preservatives.
  Step 2. At noon on Saturday, eat a normal lunch.
  Step 3. Three hours later, take two teaspoons of Epsom salts (or disodium phosphate) in about two ounces of hot water. If the taste is objectionable, drink a little citrus juice.
  Step 4. Two hours later, repeat step 3.
  Step 5. Have a grapefruit or other citrus fruits or juices for your evening meal.
  Step 6. At bedtime, you may have one of the following:
(i) 1/2 cup of unrefined olive oil followed by a small glass of grapefruit juice or
(ii) 1/2 cup of warm, unrefined olive oil blended with 1/2 cup of lemon juice.
Unrefined olive oil can be purchased from any health food store. Get the best. Using a blender to whip the oil and citrus together can improve the taste and texture substantially.
  Step 7. Following step 6, you should go immediately to bed and lie on your right side with your right knee pulled up close to your chest for 30 minutes.
  Step 8. The next morning, one hour before breakfast, take two teaspoons of Epsom salts dissolved in two ounces of hot water.
  Step 9. Be sure to continue with your normal diet and any nutritional program that has been prescribed for you, with the exception of the evening when you do step 6.

Some people report slight to moderate nausea when taking olive oil/citrus juice. The nausea should slowly disappear by the time you go to sleep. If the oil induces vomiting, you should not repeat the procedure at this time. This occurs only in rare instances. In the stool the next day, you may find small, gallstone type objects — light to dark green in color, irregularly shaped, gelatinous to cartilagenous in texture and varying in size from grape seeds to cherry pits. If you see a large number of these objects, repeat the whole liver flush in two weeks.
From New Health Perspectives, Inc. P.E.P. Personalized Education Program.

Cancer and Modern Living

As stated in the preface, cancer now strikes one in three and kills one in four Americans, about one half million a year. Since 1950, overall incidence rates have increased by forty percent. Rates have increased even more for some rare cancers.

Although the exact mechanisms are not always known, many scientists relate increasing cancer rates with the rise in use of synthetic chemicals, including pesticides, plastics, glues, etc. It should be noted that before about 1940, most Americans had little exposure to petrochemicals. Grandma's cleaning kit consisted of natural soaps, vinegar, lemon juice and other natural elements. She had no plastics or synthetic materials; clothes were made of cotton or wool; homes and their contents were made of wood or other natural materials. Small local farmers used conventionally known, natural methods of pest control and fertilized with organic wastes. Food was grown and consumed locally.

We entered the petrochemical age rapidly. As the popularity of the automobile climbed, so did uses for the petroleum distillates left over from the refining of crude oil. In 1940, we only produced one billion pounds of synthetic chemicals. In 1950, that had already risen to 50 billion pounds a year. Today, we produce at least ten times that amount, or 500 billion pounds of synthetic chemicals annually. Many of these chemicals do not break down in the environment for hundreds to thousands of years, and spread around the globe. As only one example, DDT – even though it has been banned since the 1970s – is now found in polar ice caps and in the fat tissue of fifty-five percent of the persons tested in a study by the EPA.

Most recently, a new book was released showing how man-made chemicals have an uncanny ability to mimic powerful natural hormones. These "endocrine imposters" are linked to the rise in testicular cancer and prostate cancer. These cancers appear to be estrogen-sensitive. Estrogen is mimicked by nonylphenol, a by-product of the breakdown of industrial detergents and pesticides, and added to PVC plastic.

The 1940s were also the beginning of the nuclear age. The link between radioactivity and cancer has long been known. Yet we still ignored that link for many years in a rush to prove atomic power could be used peacefully and safely. It is beginning to be more widely appreciated that nuclear power is not as safe as once thought. Even under normal operations, each plant releases thousands of curies of radioactive materials into our air and water each year. Even though there have been no new orders for nuclear power plants since 1978 or 1979, the number of nuclear plants increased from seventy-seven to 109 between 1983 and 1994. As an unfortunate consequence of the fact that no high density uses can be located near a nuclear power plant, many dairy farms are close to these plants and consequently radionucleotides can be high in commercially produced milk. Although most radionucleotides are short-lived (hours/days or months), a few have half-lives of thousands of years and become widely distributed in our environment and the food chain.

Two other factors of modern living have been linked to rising cancer rates. Strong evidence links electromagnetic fields to certain types of cancer, most notably breast cancer and leukemia in children. Diet has also been highly correlated with certain types of cancers, particulatly high red meat consumption. (See *Meats*.) It is interesting to note that indigenous peoples who eat only their native foods (and particularly no refined sugar or wheat products) do not get cancer.

Though I cannot fully endorse them without additional research, I include a few alternative practitioners whom many claim are getting miraculous results, even with advanced and "incurable" cancers. I include them both to make you aware of alternative treatment modalities and because the practitioners involved appear to be both brilliant and truly caring healers who ideas at least merit serious research.

I don't know why these researchers have not yet received more serious attention. It may be that few minds well-trained in one paradigm can be truly open to new, potentially contradictory, ways of thinking. It may also be that their widespread acceptance and use would financially disrupt the billion dollar cancer industry now governed by the current medical and pharmaceutical establishments. It may also be that paradigm shifts on a large scale take time.

One of these practitioners is Hulda Regehr Clark, Ph.D.,N.D. Dr. Clark provides exact instructions for a cancer cure in her book, *The Cure for All Cancers*. She theorizes that all cancers are caused by a single parasite, the human intestinal fluke. Exposure to propyl alcohol causes the fluke to be able to live out its entire life cycle in the body and migrate to different organs. Adult flukes in the liver of a person with cancer produces a growth factor called ortho-phospho-tyrosine. Normally, this growth factor gets produced inside a snail to help the parasite divide while reproducing. However, inside the body, it makes host cells divide as well. The book details how to clean up the environment and the body, particularly by disposing of all products containing propyl alcohol, and kill the parasite using simple herbs or a machine called The Zapper. (The herbs and the Zapper are available from the Self Health Resource Center, 800-893-1663.)

Dr. Emanuel Rivici is another practitioner whose methods have cured cancer, as well as AIDs, arthritis, drug addiction, Alzheimer's and other chronic problems. The book, *The Man Who Cures Cancer*, describes Dr. Rivici's life and work. Although his methods are totally nontoxic, he uses patented medicines.

Finally I would like to mention *The Essiac Report*, a book about "an incredibly effective Canadian herbal cancer remedy which has for more than fifty years successfully treated thousands of people suffering from various forms of terminal cancer." Dr. Charles Brusch, physician to President Kennedy, cured himself of bowel cancer using only Essiac. He has also achieved remarkable results with Essiac on cancer patients in his Cambridge, Massachusetts, clinic. The essiac formula is reputed to have Native American origins.

For information about how to obtain these books or for additional sources of information about alternate, nontoxic therapies, see *Further Resources – Books and Tapes* in the Appendix.

Aging and Modern Living

Although many books have been written on ways to slow the aging process, none is perhaps more intriguing and definitive than John Thomas' *Young Again*. The book proposes that aging and most modern degenerative and epidemic diseases are brought on by imbalances and excesses in the body due to toxic chemicals and other stresses of modern living. Detoxify the body, give it plenty of pure, "living" water and air, healthy raw or slightly cooked foods (organic or home grown on "living" soils), some key super foods and supplements, regular routines and plenty of rest and exercise, and most disease can be reversed. John Thomas explains detailed physiological pathways whereby cancer, arthritis, diabetes, hardening of the arteries, balding, HIV, obesity, edema, chronic fatigue and many other diseases can be reversed once the physiology is functioning as it should. He believes that we are all subclinically ill. When a full blown disease appears, the process of aging is already far along. If one detoxifies, time appears to slow, and there is no reason we might not live for hundreds of years.

The book is not shy about flying in the face of conventional wisdom, even concerning some sacred cows like tofu and alfalfa sprouts. Conventional scientists and health practitioners will find some of its conceptual constructs beyond the pale of believability. Moreover, his monolithic picture concerning the current medical/pharmaceutical/industrial powers may not be particularly constructive or even entirely accurate. However, he is a thinking individual who researches thoroughly before he forms an opinion. His ideas are well aligned with the wisdom of the ancient Ayurvedic physicians of India, certain indigenous cultures and even theoretical physicists who are beginning to marry modern physics with ancient mysticism. Finally, his own health and youthful vitality are evident from his photo and testify to his formulas. At age fifty-one, he appears to be in his thirties. According to his text, a physician who performed his hernia operation commented that Thomas had the organs of a fifteen-year-old.

The kernels of his formulas are summarized here. The "don'ts" are usually tied to specific ailments. The recommendations (or "dos") to maintain perpetual youth and stay free from disease, however, are basically consistent for every complaint.

Recommendations (Dos):
1. Detoxify your body. Depending on your level of health, this is more or less urgent. Detoxify slowly and don't expect to feel better right away. In fact, you may feel much worse for awhile as the toxins that have been bound in fat tissues are released and flushed from your system.
a. Prepare to detoxify with Yucca Herbal Blend™ and proanthocyanidins (PACs).
b. Flush your liver of stones. (See *Liver and Gallbladder Detoxification.*)
c. Receive a colonic.
d. Drink one 8 oz. cup of purified BEV™ or Medical Grade Ionized water every hour. (See explanations in section on *Water Purification.*)
2. Use regular routines. Rise, eat and sleep at regular times in rhythm with nature.
3. Don't eat between meals. Give your digestive system a rest. Never eat large meals right before bed. Do occasional therapeutic fasting.

4. Do daily aerobic exercise. He recommends rebounding, meaning jumping and swinging the arms on a small trampoline, at least twelve minutes a day.
5. Massage the lymph system daily.
6. Eat lots of fresh fruits and vegetables, preferably from your own organic garden. At least one third should be raw, and the remainder lightly steamed or stir-fryed. Don't drink during a meal.
7. Get plenty of organic potassium and B vitamins from fresh greens, including comfrey, kale, collard greens and tobacco you grow yourself. (YES TOBACCO, but make sure it is the edible variety. Eat with apple cider vinegar/fresh lemon.)
8. Eat real proteins: lentils, beans, legumes (except soy), clean organic fish, fowl and meat.
9. The best oil is virgin olive. Sesame, sunflower and flax seed are also good.
10. Drink only purified water using reverse osmosis and the BEV™ system. Drink plenty of water every day. One eight ounce glass every hour is best.
11. Use a shower filter or home system to filter out chlorine. End the shower with cold water to stimulate the production of brown fat. (A good type of body fat.)
12. Substitute Bragg's liquid aminos or liquid ionic sea minerals for salt.
13. Spend time in nature and in meditation. Walking barefoot on the grass or in the sand is therapeutic.
14. Replace mercury fillings with epoxy plastic with a qualified holistic dentist.
15. Take certain super foods, teas, and supplements on a regular basis. These include Yucca extract or Yucca Herbal Blend™, Kombucha tea, Rene's tea, proanthocyanidins (PACs), enhanced trivalent chromium (ETVC), Gerivital, super blue green algae, Harmonic pollen from British Colombia and capsules of fruit juices, vegetable juices, and biogenic liver capsules from Argentina.
16. Use an ozone generator and ionizer. (See *Negative Ions* and *Air Purification*.)
17. Use something to handle electromagnetic fields. He recommends Biomagnets.
18. Sleep outside.
19. Use homeopathic medicines. Take certain homeopathics to neutralize the effects of various vaccines you have had in your lifetime (polio, hepatitis, tetanus).
20. Think positive thoughts. (See *Inner Housecleaning*, pg. 80.)

The Don'ts:
1. Don't drink any kind of soda due to sodium and aluminum content, or alcohol.
2. Don't eat sugar, cane syrups or aspartame (NutraSweet).
3. Eliminate fluoride from your diet. Don't use fluoride toothpaste, and use reverse osmosis to take fluoride out of your drinking water.
4. Don't drink from public water supplies and don't buy bottled water in plastic, particularly distilled water as it is 'dead' water.
5. Eliminate any potential sources of aluminum from your diet – no sodas in aluminum cans, no aluminum cookware, no baking powder with alum.
6. Don't use deodorant (particularly with aluminum). After detoxification, your body odor should have no smell, like the bodies of children.
7. Stay away as much as possible from prescription medicines.
8. Stay away from irradiated, genetically engineered or industrially grown food.
9. Don't eat salt and processed foods high in sodium!
10. Don't eat any soy product; or margarine, canola or soy oil. (See *Oils*.)
11. Don't take any kind of artificially synthesized vitamins.
Source: Thomas, J., 1995. *Young AGAIN!* (See *Further Resources – Books*.)

Air and Air Purification

A growing number of people are falling victim to "sick house syndrome." Most people do not even relate their symptoms to indoor air pollution, but consider these details:

o Toxic chemicals outgas from wallpapers, paints, varnishes, new furniture, drapes and carpets, synthetic fabrics and plastics; even toys, tap water and cleaning agents as described in the previous chapters.

o Concentrations of about twenty toxic compounds were usually ten to twenty times higher, and as much as 200 times higher, inside homes than outside homes, according to a five-year study by a Harvard scientist for the EPA. It didn't matter whether the homes were in rural areas or next to a chemical factory.

o Nearly 500 harmful chemicals were found by the EPA in a 1988 study of ten buildings over a five-year period.

o The EPA has called indoor air quality "the most significant environmental issue we have to face now."

o Our homes and office buildings are now so full of synthetic chemicals that toxic smoke, not fire itself, is the cause of eighty percent of fire related deaths.

o Of nearly 12,000 homes sampled in 1986 by the EPA, one in five had radon levels exceeding the EPA action level. Radon is a radioactive gas causing cancer.

There are four basic ways to treat air mechanically that has unhealthy chemicals, particles or offensive odors: (i) filtration; (ii) purification by chemical means, including adsorption, a combination of chemical purification with filtration; (iii) ionization, which causes particles to settle out of the air by attaching a charge to them, including electrostatic precipitation, a combination of charging particles and filtration; and (iv) coverage, or masking the offensive odor(s) with another stronger, but more pleasant odor. The fourth method is not recommended when the pollutant is harmful to health.

Different types of filters have different capacities for the size and type of particles captured. The most effective particle filtration method known is High Efficiency Particulate Air (HEPA) filtration. Computer chip manufacturing, pharmaceutical production and hospital operating rooms, where totally clean air environments are required, all use this method. A true HEPA filter is made of special fibrous material, featuring an extended surface that is deeply embedded with randomly-positioned glass fibers. A true HEPA filter media is certified to have a minimum efficiency of 99.97 percent in capturing particles of 0.3 microns in size. This size gains easiest entry into the respiratory system and places the greatest burden on the body's defense system. It is also the most difficult respirable size particle to remove from the air. To give you an idea of size, the dot in the letter "i" is 500 microns across. Other media filters are typically between twenty percent and eighty percent efficient.

Adsorbents are porous materials full of tiny pores through which gases can pass. Common adsorbents like activated charcoal, silica gel and activated alumina react with contaminant molecules, causing them to cling to pore walls. In this way, adsorbents differ from filters which simply trap particles. (The exception would be electrostatic precipitation, described below.) Adsorbents remove gases well, but usually only large gas molecules, such as formaldehyde and ammonia. Even though thousands of tiny pores give these materials a large surface area, eventually all the pore sites become full and the material needs to be replaced. The best filters nowadays have both HEPA and adsorbent materials.

Air purification through chemical means is most commonly accomplished with ozone. Most of us know ozone as the substance in the upper atmosphere that protects us from too much ultraviolet radiation. (See *General Environmental Issues*.) We also associate ozone with smog in cities. It is not commonly known that ozone is generated by a thunderstorm and purifies air naturally. Ozone (O_3) is an oxygen molecule (O_2) with an extra oxygen atom. It is highly reactive and breaks down many pollutants into safe-to-breathe products such as water (H_2O) and carbon dioxide (CO_2). Ozone also can help destroy mold, mildew, fungi and bacteria. Ozone can now be artificially generated to purify air. It is particularly effective for smoke fumes and is used by cleaning companies to rid fire-damaged sites of offensive smoke odors. There is controversy about whether ozone is appropriate for use in the home environment. Ozone in higher concentrations is a respiratory irritant, and can be dangerous to those with pre-existing respiratory or heart conditions. Other research suggests that low levels of ozone may actually be beneficial for health and ozone is used in various therapies. The Federal Clean Air Act sets a limit of .12 ppm outside, and the FDA sets a limit of .05 ppm for the ozone from electronic air cleaners. If you purchase a unit, make sure that it can perform to this standard and below .05 ppm. Some air ozonators contain a device to shut down ozone generation if levels rise to this threshold.

Negative ions are generated naturally during thunderstorms, or by water crashing against rocks, as it does at the ocean shore or waterfalls. Negative ions provide another natural means to clean air. Negative or positive ions attach themselves to floating air particles. These particles are removed from the air when they attach to an oppositely charged surface or attract enough dust to drop from the air due to weight. The use of a negative charge is particularly effective over time for removing the smallest of particles or gas contaminants from the air. We now can artificially generate ions for air conditioning and purification purposes. (See *Negative Ions and Health*.) Negative ion generation is effective for minimizing dust in dust-producing industries such as wood-working shops, and is more effective in removing cigarette smoke than the typical fan/filter type of air purifier.

Electrostatic precipitation charges airborne particles with a negative or positive charge. Once charged, special positively or negatively charged collector materials in the cleaner trap the particles. Electrostatic filters are now produced to replace regular furnace filters. They can do a reasonable job in minimizing dust, smoke particles and some allergens. Like adsorbents, their effectiveness drops as sites become filled and they should be kept clean. HEPA filters are also manufactured for the furnace, but furnace motors should be sufficiently large to handle the strain.

Negative Ions and Health

The atmosphere always has a certain number of particles in the air with negative and positive charges. The relatively clean air of open country has 2,000 to 3,000 charged particles or ions per cubic centimeter of air. Around waterfalls, health spas and after thunderstorms, this figure can reach as high as 35,000 to 100,000.

Many believe negative ions are important to the optimal health of living things. Although the exact mechanism is not well understood, some people appear so sensitive to an imbalance in the concentration of ions, particularly a lack of negative ions, that they suffer severe nausea, migraines, unexplained anxiety, depression and depression-related symptoms, like insomnia and lack of appetite.

Unfortunately, city life and air conditioning deplete the air of negative ions. Concentrations in air in cities, cars and in heavily air-conditioned buildings can drop as low as twenty to 200. Certain places in the world seem to suffer unexplained drops in the concentration of negative ions at certain times of year. During these episodes, people report rises in accidents, crimes, suicides and doctor visits. Sensitive people living in these areas suffer particularly. When the concentration returns to normal, or when a sensitive person leaves the area, his or her symptoms immediately lift.

Ion research as been conducted since the 1920s. Some modern research has had conflicting outcomes and controversy regarding the topic. The phenomena reported above, however, have been extensively researched by scientists in different parts of the world, including Dr. Felix Sulman of Hebrew University in Israel, Dr. Albert Krueger at the University of California at Berkeley, A.L. Tchjijewsky and A.A. Minkh of Russia and Christian Bach of Denmark. Recent reports giving evidence of the positive effects of exposure to high doses of negative ions have been provided by former Societ Union researchers and reported in Muscle and Fitness (Oct. 1993).

It was only in the 1950s and 1960s that we began using generators commercially to increase the concentration of negative ions in the air to improve our health and well-being. The early negative ion generators shot out a stream of negative ions in a beam from the unit. The beams could not reach very far from the source, and people had problems with "black wall," the deposit of negatively charged particles on a positively charged wall. In 1986, a new generation of ion generators became available. These generators create a proper balance of negative and positive ions (3,000 positive ions and 4,000 negative ions per cubic centimeter) in an area sixty feet in all directions from the unit, even through walls and floors. Since both negative and positive ions are generated, the black wall problem disappears.

The Ion Effect by Fred Soyka with Alan Edmonds is a good reference book in support of the phenomena. Bantam Books published it in 1977, but it is no longer for sale. Try to find it at your local library. (See *Further Resources.*)

Water and Water Purification

According to Carol Browner, head of the Environmental Protection Agency (EPA), most households receive safe drinking water. However, a recent EPA report shows that in 1994 about one in ten Americans — or some 30 million people — were served by drinking-water systems that violated public health standards. Many of the violations involved bacterial contamination that caused acute, short term illness. Stories of more serious problems still crop up from time to time, and many authorities believe that the problem is more widespread than we know and growing. Some examples from various sources verify this perception:

o The residents of Washington farm country, an area with pristine emerald forests, began to experience cancer, the death of pets and chronic illness among their children. Eventually they found that their well water was contaminated with dichloroethylene and trichloroethylene (TCE). A former waste dump at the McChord Air Force Base was suspected, although officials denied responsibility.

o In Woodstock, New York, the county health department tells residents not to drink or cook with their water because it is highly contaminated with asbestos leaching from old water pipes.

o In Pittsfield, Massachusetts, an estimated 8,000 people fell ill with diarrhea when the parasite giardia got into the town water supply.

o In East Gray, a small, bucolic town in Maine, residents began to get unexplainable skin rashes. The children were always sick, the adults were tired and headachy, and the town and area residents had an unexplainably high rate of miscarriages and low-birth-weight babies. When someone finally suspected the water, two highly toxic chemicals were found: trichloroethylene and tri-chloroethane. The source was located in hundreds of rusted barrels in a gravel pit, with others scattered nearby, all leaking chemicals into the ground.

o In South Carolina, near Myrtle Beach, unusually high levels of sodium, which has been linked to heart disease and high blood pressure, were discovered in public water.

o In Milwaukee, an outbreak of crytosporidium in 1993 sickened 403,000 people with stomach upsets and diarrhea, and 100 people died.

Water contains many possible types of contaminants. These include pathogens like bacteria and viruses; and protozoa, like cryptosporidium and giardia. Water may also have radioactive contamination. According to the EPA, about 50 million Americans are exposed to the odorless, tasteless gas radon in their drinking and bathing water. Radon is a proven cause of lung and rectal cancer. Various inorganic elements, such as sodium, lead or arsenic, are also sometimes found in water. Lead most often comes from lead pipes or solder in our own plumbing. Lead impairs children's attention span and IQ levels, and causes high blood pressure in adult men. The EPA estimates that about one in six people drinks water with excessive lead.

61

Next to lead, nitrate is the most serious inorganic contaminant. Nitrate contamination occurs in groundwater and mainly concerns people with private wells in rural areas. Groundwater contamination with nitrate stems from the use of chemical fertilizer on farmland and water leaching from animal feedlots and septic tanks. Nitrate poisoning is particularly harmful to infants and children, and in severe cases, can result in brain damage and death.

Organic compounds constitute another type of water contaminant. Organic compounds contain a carbon atom, the basic element in the bodies of most life forms. Organic contaminants include pesticides, gasoline, gasoline byproducts and industrial chemicals, particularly those made from petrochemicals. Many compounds made from petrochemicals have been found to be either carcinogenic (causing cancer), mutagenic (causing genetic mutations), or teratogenic (causing birth defects). New research has linked certain organic compounds to interference with our endocrine system. These endocrine imposters mimic our natural hormones – hormones like those that regulate sperm production, our immune system and cell division in the developing human brain.

Chlorine, a chemical used to disinfect our water, has been linked to hardening of the arteries, fatal heart attacks and bladder and rectal cancer. In addition, chlorine reacts with organic matter in our water, such as decaying leaves, to form cancer causing trihalomethanes (THMs). In Washington, D.C., in summer, for example, THMs can reach levels thirty percent above the health limit when warm weather forces the city to use more chlorine to kill bacteria.

With the proliferation of water purification systems and claims, as a consumer you may find it difficult to make an educated decision. Because it can be expensive to try to solve a problem that you don't have, you want to have an idea of the potential contaminants in your water that need to be removed. Talk to your local water authority and ask for their testing results. Remember, however, that these results are a yearly average, and your water might not comply at any particular point in the year. Speak to local environmental groups and environmental reporters and conduct your own library and newspaper research. Finally, you may want to have your water tested by an accredited lab. (See *Further Resources.*)

Different types of contamination require different treatment. Water can be treated or purified through boiling, mechanical filtration, adsorption, aeration, the addition of gases or chemicals and ultraviolet light. Boiling, if done for at least fifteen minutes, can kill most bacteria and other parasites. It will also boil off chlorine and volatile organic compounds (VOCs, See *Toxic Substances and Health Effects*), but heavy metals and inorganics like lead will remain in the pan. Distillation uses boiling to purify water, but has the opposite end result. Steam rises and condenses and leaves most contaminants behind. Distillation will kill bacteria and is effective with heavy metals and inorganics, but volatile organic chemicals with a boiling point lower or equivalent to water will condense with the steam and remain in your water supply. Distillation is also slow, typically five hours for one gallon.

Bacteria, viruses and parasites are some of the more difficult water contaminants. Boiling, as stated earlier, can kill most disease causing bacteria, but some viruses and parasites are not destroyed with boiling. Most large municipal water companies use chlorine to kill bacteria. This system is not 100 percent effective in killing all bacteria, but brings levels down to a point where most healthy persons are not affected. Chlorine does not kill many viruses or some parasites, e.g. cryptosporidium. Ozone gas is also a powerful disinfectant. Its popularity is growing, especially in hot tub applications. Ultraviolet light is one of the most effective means of killing bacteria and viruses – killing 99.9 percent. However, it does not kill some encysting parasites like crypotosporidium and giardia. A system that filters down to one micron or smaller, such as reverse osmosis, carbon or porcelain block, or pre-coat systems, can help with these. However, no mechanical filtration system alone can guarantee complete removal of bacteria and viruses as some of these organisms are extremely small. Most filtration systems (except those with UV light) are made for use with potable (already germ-free) water. Care needs to be taken, especially in bed type filters, that the filter itself does not become a breeding ground for bacteria. Silver nitrate is sometimes added to bed type filters to try to inhibit bacterial growth, but the silver nitrate itself is a toxic substance.

Carbon filters work through adsorption, or the adhering of contaminants to the walls or pores of carbon media. Carbon is the best means of removing chlorine and organic chemicals like pesticides and industrial and petrochemicals. It also removes bad tastes and odors. Carbon filters are generally made from granulated charcoal from burning coconut shells. Carbon's filtering capacity increases with surface area. Therefore, the more finely ground or powdered the charcoal, the more surface area will be available. Most carbon filters are bed type filters made with granular activated carbon (GAC). The only potential problem with bed filters is a tendency to channel, or form passages where water can pass through unfiltered. Carbon block filters or carbon pre-coat filters, both of which filter down to one half micron, are more reliable. With a pre-coat filter, the carbon is microscopically fine, greatly increasing surface area and therefore filtering capacity.

Spool-type filters, which look like tightly wound string or yarn, mechanically filter sediments. These filters are used in photographic labs, where it is important to remove large particles from photochemical solutions. They are also used as components in other types of systems, such as reverse osmosis or carbon block, to protect the filter from becoming clogged too rapidly. Sudden changes in water pressure, however, can cause these filters to flex and dump their sediment loads.

Carafe type filters also depend entirely on mechanical filtration. Water poured into the top compartment of the carafe trickles through the filter and collects in the pitcher below. Carafe filters are better than no filtration, as they can remove about fifty percent of toxic, volatile organic chemicals, sixty-five percent of bacteria, and from fifty to eighty-five percent of lead, with a fresh filter. Carafe filters are not an acceptable form of purification where lead or other contamination is severe. Moreover, they can cost as much or more to operate than a reverse osmosis filter. At $7 a filter, a family of four should use twenty-five filters a year or spend $175.

Reverse osmosis is currently considered the cadillac of point-of-use water treatment. A good reverse osmosis system removes ninety to ninety-nine percent of all contaminants. They also remove inorganic minerals. Minerals are best obtained from good food, in any case, or in pill supplements bound with enzymes or vitamins from a food source. Reverse osmosis uses a semipermeable membrane that allows water and small organic molecules to pass through, but excludes ions and large molecules. It effectively removes inorganic contaminants like lead, arsenic, fluoride and nitrate, but usually allows volatile organic chemicals (VOCs) to pass through. To remove these, reverse osmosis systems include one or more carbon filters as well, and a sediment pre-filter. Reverse osmosis generally produces three to five gallons of waste water for every gallon of purified water. It is important to have a system that automatically shuts off when the storage tank is full.

If the water supply in your home is contaminated with radon, you have a special case. Radon in water is released into the air when you run the water, as in a shower. Therefore, radon contamination requires treatment of all the water entering the home. (See *Radon*, Chapter I.D. and *Water*, Chapter II.F.) Radon is a serious issue. According to one EPA official, radon alone may cause more cancer deaths than all other water contaminants combined. Generally, levels in your air will be about one tenth that in your water. According to the EPA, you should take action to reduce radon in your water only if it is tested to be 10,000 picocuries per liter or higher. Have your water tested only if your indoor air radon level is high, and if you get your water from a private well or small community water system. (See *Further Resources, Radon* and *Testing*.) The removal of radon requires a whole house carbon system or a home aerator. Aerators cause radon to bubble out of the water by agitating it with pumped-in air. The radon is vented outside with a pipe. Aerators are generally more expensive than carbon systems.

A whole-house carbon system may be a good idea in any case. Carbon removes chlorine and various volatile petrochemicals that are not healthy to breathe or bathe in. Removal of these types of contaminants can often clear up skin or scalp problems, and prevent the absorption of potentially toxic or hazardous substances through the skin. Persons with immune system disorders might also consider installation of a whole-house ultraviolet (UV) system, as bacteria can be abosrbed by the skin as well. (See *Water*, Chapter II.F.)

As more than eighty-five percent of the U.S. has hard water, many people have a water softener installed for the whole house. Softeners remove calcium and magnesium through the use of synthetic resin beads. The resin beads must be regenerated every now and again with salt water from an adjacent brine tank. Softeners help eliminate mineral deposits in tubs and sinks, the build up of scale in hot-water heaters and other appliances, spots on dishes and dull looking laundry. People with soft water also need less soap, their food tastes better and they have softer skin and shinier hair. Soft water also saves on utility bills. Tests show that water heaters run twenty-two to thirty percent more efficiently with soft water, and dish and clothes washers can be run with cooler water and shorter cycles. Unfortunately, softener salts are polluted with strontium, aluminum and chromate. The salts can corrode old galvanized pipe, accelerating the seepage of cadmium into the water. Use of a softener, then, necessitates the use of a reverse osmosis point-of-use system for drinking and cooking water. One author prefers conversion of the house to PVC pipes to eliminate hardness problems. (See *Water*, Chapter II.F.)

In my research, I discovered several esoteric forms of water treatment, which I will attempt to describe. These techniques are designed to restore the "life force" to water and to remove negative imprints in water resulting from pollution. The treatments are on the fringe of science, but I include them for your interest and because of the potential validity of their theoretical constructs and the processes that stem from them.

Based on the discoveries of quantum physics, science is currently moving through large and continuous paradigm shifts which have profound implications on what we consider rational and possible. These technologies appear to work on subtle energetic levels and their soundness is therefore not as simple to assess as technologies based on classical physics. If you are comfortable with forms of self-diagnosis like muscle testing, you can perform your own evaluation to see if the treatments will benefit you. They include the use of Coral-CalciumTM, the use of magnetic fields or the Grander system, the BEVTM system and a MikrowaterTM treatment that comes out of Japan.

Coral-CalciumTM was discovered by a reporter who went to Tocunoshima, Japan, to interview the world's oldest living man, born in 1864. The reporter discovered that most people in the area lived to be over ninety-five and were free of degenerative disease. The only common factor seemed to be the water from a water table enriched with large amounts of coral calcium from the erosion of the coral reefs surrounding the island. When others drank water treated with Coral-CalciumTM, they reported remission of problems and symptoms associated with arthritis, high blood pressure, angina and clotted blood vessels, skin problems and even cancer. No one understands completely how Coral-CalciumTM works, but it does appear to raise the pH level in water and the body, making it more alkaline and lowering electrical potential. One source states that the substance works by transferring earth energy to the water. This cancels low energies and enlivens the molecules so that they vibrate at a frequency that promotes healing.

Some scientific studies conclude that water can indeed carry an electromagnetic footprint from pollution. Dr. Wolfgang Ludwig, Ph.D. (Physics and Natural Sciences) found a certain unfavorable frequency of electromagnetic radiation that corresponds to harmful substances like cadmium, lead, nitrates and bacteria. Even when one of these substances is removed, the water retains these signals. It is as if the electromagnetic oscillations have been transferred to the water, and the water "remembers." Similarly, the water can remember life-enhancing frequencies. Contaminating water, forcing it through pipes, even distilling it, all take the life force out of water; or, in other words, give it unhealthy, rather than healthy electromagnetic frequencies. In nature, water can apparently dump its contaminant load, and even regain healthy frequencies through the natural, hydrological cycle of seeping into the ground, or flowing through rivers to the ocean, evaporating through solar radiation into clouds, and eventually coming down again as rain. Agitating water – which occurs naturally when water flows down a mountain –

provides one simple method to help erase damaging frequencies. Methods to use wireless magnetic fields to transfer the vibrations of health to water are now being studied. Methods to restore healthy frequencies through the use of magnets have been patented since 1950. Some studies indicate that magnetized water positively affects everything from the deposit of scale on pipes to the health and productivity of farm animals.

The BEVTM system sounds similar to magnetized water, but supposedly goes beyond it. B.E.V. (BEV) stands for Bio-electric Vincent in honor of professor Louis-Claude Vincent, who developed the standards for biologically friendly water. The BEVTM system reprograms the electromagnetic signature of water so that it is completely body-friendly, and removes the frequencies of contaminants that are unfriendly to the body. The BEVTM system removes regular contaminants by pre-stages virtually identical to those of reverse osmosis. According to promoters, the last or B.E.V. stage, produces water that "reprograms the electrical rhythm of the body and awakens natural harmonic frequencies at the cellular level and below." (Thomas, 1995)

Mikrowater is a special type of therapeutic water being used in Japan to treat everything from diabetes and cancer to gout and connective tissue disorders. Mikrowater is made using ionizers that can alter the pH of water from highly acid to highly alkaline depending on the condition being treated. (Sick bodies are described as acid bodies, and alkaline water taken internally promotes healing. Acid Mikrowater is used externally to heal infections and promote beautiful skin.) Mikrowater ionizers are also supposed to "create extremely fine water molecules that powerfully interact with body tissues at the subtle energy level where disease and illness have their roots." When ionic sea minerals are added to BEVTM water and used in a Mikrowater ionizer, water with a superior healing potential gets produced (Super Medical Grade Water). (Thomas, 1995)

Electromagnetic Fields (EMFs)

The subject of electromagnetic fields and their effects is beginning to creep more and more into our awareness. Several books have been published on the topic (See *Further Resources*) and both research and popular literature are giving it greater attention. For example, *New Scientist* (October 1995) reports: "Millions of people may face an increased risk of cancer and degenerative diseases because they are exposed to electromagnetic radiation from power lines and household electric appliances. The warning comes in a leaked report prepared for the U.S. government's radiation advisers, the National Council on Radiation Protection. . . The report, a draft compiled over the past nine years by a committee of eleven, leading American experts in EMFs . . . is the most comprehensive study ever on the health effects of low-frequency EMFs. The committee's chairman, Ross Adey, . . . says there is now a 'powerful body of evidence' to suggest that very low exposures to EMFs have subtle, long-term effects on human health."

Although these broad statements remain controversial, and some studies produce conflicting results (see 1996 National Science Foundation report), it has been well accepted that pregnant women at least should not spend long hours in front of a computer screen, and that people whose careers put them in front of a screen eight hours a day should take steps to protect themselves. Until now, the placement of shielding or blocking mechanisms in front of the computer screen was the best that could be done. Some computer companies have also begun to manufacture computers that reduce radiation exposure. Although our proximity to computers makes them of particular concern, we truly live in a sea of electromagnetic radiation from all sorts of electric appliances, the electric wires running in our walls and even a city's power grid.

The ubiquitous nature of electomagnetic radiation, and our total dependence on electricity, makes everyone nervous about the implications of negative findings about EMFs. Moreover, we only know two ways to protect ourselves: to block or distance ourselves from the fields.

New principles discovered in quantum physics suggest that it is not the intensity of the fields so much as their incoherence that produces harmful effects on human physiology and brain functioning. This understanding has enabled scientists to produce a simple breakthrough technology, called UT code, to change the random, chaotic motion of electrons, electronic "noise," into coherent, harmonious waves. Electromagnetic fields that surround all electric devices become more orderly. This seems to alleviate the many symptoms of everyday exposure to electromagnetic fields. Due to this technology, we can now enjoy the benefits of life-enhancing electrical products in greater health and comfort. (Verified at Dr. Robert Perry and Associates, an independent research firm in Fallbrook, CA.)

Some people wear a neodymium magnet or a BioElectric Shield, which uses specially altered quartz crystals, to balance energies and help cope with the frequencies of the alternating current in use today to power appliances and homes. (For more information on where to find EMF products, see *Product Sources, Electromagnetic Field Coherence.*)

Health Effects of Toxic Substances

Artificial or Synthetic Colors or Dyes. Artificial dyes can be found in every type of processed foods, cleaning products and personal care products, including cosmetics. They will be labeled as FD&C or D&C, followed by a color and a number. Example: FD&C Red. No. 4, or FD&C Yellow No. 6. Most FD&C (food, drug and cosmetic) and D&C (drug and cosmetic) colors are made from coal tar, which has been shown to cause cancer in animal tests. Other families of synthetic dyes may contain or be made from phenol, formaldehyde, carbon tetrachloride, benzene and xanthenes (used in lipsticks), among other toxins and carcinogens. There are many natural alternatives to synthetic dyes, including indigo, henna, beet powder, caramel and so on.

Asbestos. Asbestos is the collective term used to describe several silicate mineral fibers that tend to be flexible, incombustible, durable, and good thermal and electrical insulators. Because of these excellent qualities and its cheap availability, it was used extensively in all types of construction and in home appliances until about 1960 when its long term health effects began to be known. As long as the sharp fibers remain in place, asbestos is relatively safe. If released from its binding material through erosion, vibration or renovation, and allowed to become airborne, it poses a danger to health. Those fibers 0.1 to five microns in size are readily inhaled and lodge in the lung and lymph ganglions, where they remain permanently, causing asbestosis, tumorous growths, fibrosis, and even lung cancer.

Carbon Monoxide. Carbon monoxide (CO) is a colorless, odorless, toxic gas formed by the incomplete combustion of fossil fuels. It forms in automobiles which are left running; by wood- or coal-burning stoves and oil furnaces; in poorly tuned gas furnaces, ranges and ovens; and cigarette smoke. It is the most prevalent and potentially dangerous of all indoor pollutants. CO interferes with the blood's ability to transfer oxygen throughout the body and produces headaches, impaired vision and judgement, irregular heartbeat and exacerbation of cardiovascular symptoms. Chronic exposure can be misdiagnosed. Exposure to 1,500 ppm for one hour is usually fatal.

Chlorine. Exposure to chlorine may produce pain and swelling of the mouth, throat and stomach; erosion of mucous membranes; severe respiratory tract irritation; pulmonary edema; skin eruptions; vomiting; circulatory collapse; confusion, delirium and coma. Chronic exposure has been linked to high blood pressure, anemia, diabetes, heart disease and a forty-four percent greater risk of gastrointestinal or urinary tract cancer. According to clinical observation by medical doctors, chlorine fumes rising from hot or cold running tap water while taking a shower or washing the dishes can cause red eyes, sneezing, skin rashes and dizziness or fainting.

Fluoride. Although more controversial in the U.S., fluoride has been banned in ten European cities due to linkages to cancer and genetic damage. Fluoride may cause tiredness, weakness and many other skin, digestive and nervous disorders. Excessive fluoride can reduce blood vitamin C levels and weaken immune system function.

Formaldehyde. Formaldehyde is a colorless, organic compound with a pungent odor that can be smelled at one part per million or less. It is extensively used in wood products such as fiberboard, particleboard, and plywood, and permanent press clothes, bedclothes and drapes. Formaldehyde is highly irritating to the eyes, skin and mucous membranes, causing burning and tearing eyes, irritation of the nose and throat, contact dermititis, hives, coughing and even a bronchial asthma attack. It can easily enter the bloodstream, potentially producing headaches, drowsiness, nausea, vomiting and diarrhea. Studies have linked nasal cancers and impairment of nasal membranes in rats and mice to prolonged formaldehyde exposure.

Lead. There is no demonstrably safe level for lead. Unfortunately, because of the widespread use of leaded gasoline, lead-based paints and other such lead based products in the past, lead exposures in air, water and food can only be minimized, not avoided. Chronic low-level exposure can affect the heart, immune system, nervous system, the kidneys, liver, gastrointestinal tract, the bloodforming system and the reproductive system (causing malformations of sperm and low sperm counts). Chronic low-level exposure in children produces low IQ, short attention span, hyperactive behavior and motor difficulties. Higher exposures can lead to early symptoms of lead poisoning, including abdominal pains, loss of appetite, constipation, muscle pains, irritability, metallic taste in the mouth, excessive thirst, nausea and vomiting, muscle weakness, headache, insomnia, depression, lethargy and shock.

Nitrogen Oxides. Nitrogen oxide (NO) and nitrogen dioxide (NO2) are produced by burning natural gas or oil in oxygen-rich environments. Gas stoves and ovens provide a primary source, but other sources include furnaces, unvented gas and kerosene heaters and cigarette smoke. These compounds constitite respiratory irritants at low concentrations and can cause lung damage and even death at high concentrations (above 150 ppm). Lower concentrations can produce chronic lung disease and bronchitis. Some studies have found a correlation between an extensive history of respiratory ailments in children and the use of gas versus electric stoves.

Pathogens. Potential disease-producing agents such as bacteria, fungi, molds and viruses abound in offices and homes. Each human sneeze releases up to 40,000 microbe-filled droplets. Each of the millions of human skin scales we shed daily carries an average of four bacteria. Dog and cat dander carry their own host of pathogenic organisms. Rugs and carpets are particularly good hiding places for millions of microbes.

Petroleum Distillates. The refining of crude oil into gasoline leaves behind a wealth of chemicals that as a group are called petroleum distillates. The distillation of oil into ethanol leaves many volatile organic compounds that do not burn as well as gas. These chemicals have been the feedstock for the petrochemical and plastic age. Most consumable products now use some type of petrochemical in their formulation. Even snythetic clothes and plastics have a petrochemical basis. Many of these chemicals are now being found to cause cancer or otherwise stress living systems, and some are the most toxic and carginogenic chemicals known. (See Volatile Organic Chemicals, VOCs.)

Polychlorinated Biphenyls (PCBs). PCBs are nonflammable, chemically and thermally stable, organic compounds, used widely as electrical insulators in the U.S. for fifty years. They were also used in textile dyes, printing inks, paints and carbonless copy paper. You might find them in the electronic ballasts of fluorescent light fixtures manufactured before 1978, and a burned out ballast can raise room levels fifty times higher than normal. In the mid-1960s, PCBs were discovered to be a persistent and motile contaminant, found everywhere from the Antarctic ice to mother's milk. PCBs are highly toxic substances that can result in death in high enough concentrations and are also carcinogenic, mutagenic and teratogenic. They can build up in living tissues, creating a cumulative effect, and will also cross the placenta to the fetus in pregnant women.

Radon. Radon 222 is a radioactive gas that occurs naturally in many kinds of rocks and soils. It has also been found in building materials such as concrete and brick made from this radioactive rock. Radon can migrate from the material it is in and decays into radioactive "daughter" elements – polonium 218, lead 214, bismuth 214 and polonium 214 – which emit significant alpha radiation. These electrically-charged daughters attach themselves to respirable particles floating in the air and are then inhaled. EPA researchers blame indoor exposure to radon progeny for most non-smoker, lung-cancer deaths – an additional 10,000 to 20,000 per year.

Volatile Organic Chemicals (VOCs). Carbon forms a major part of the chemical makeup of organic chemicals. VOCs are organic chemicals that vaporize readily at low temperatures. They include the plentiful halocarbons (ringed carbons) and hydrocarbons (carbon-hydrogen compounds) found in petroleum distillates. Studies conducted in the late '80s by Harvard for the EPA and the Consumer Product Safety Commission (CPSC) found large numbers of such chemicals (between twenty and 150) in test homes at levels five to ten times greater than those found outdoors. VOC sources in the home included aerosols, cleaners, paints, plastics, pressed wood products, furnishings and pesticides, among others. Automobile upholstery can release more than 147 different organic compounds, including the highly toxic vinyl chloride, within the car's tightly enclosed space. The vast number of VOCs makes it impossible to describe their health effects as a whole, but they are commonly (i) carcinogenic (causing cancer); (ii) teratogenic (causing birth defects); (iii) mutagenic (causing genetic mutations); (iv) irritating or damaging to the mucous membranes or respiratory system; (v) damaging or depressing to the brain or central nervous system and (vi) damaging to the liver, kidneys, red blood cells or bone.

Cosmetic Ingredients to Avoid

From Aubrey Organics ® "Natural Ingredients Dictionary" by Aubrey Hampton.

Diazolidinyl Urea and **Imidazolidinyl Urea.** These represent the most commonly used preservatives after the parabens. They are well established as a primary cause of contact dermititis (American Academy of Dermatology). Two trade names are Germall II and Germall 115. Germall 115 releases formaldehyde at just over 10o. These chemicals are toxic.

Butyl/Ethyl/Methyl/and Propyl Paraben. Manufacturers use these as inhibitors of microbial growth and to extend shelf life of products. They are widely used though they are known to be toxic, and cause allergic reactions and skin rashes. Methyl paraben combines benzoic acid with the methyl group of chemicals. Highly toxic.

Petrolatum. Ironically, petrolatum is mineral oil jelly, and mineral oil causes the same skin problems that it is advertised to alleviate. It leads to dry skin and chapping by interfering with the body's own natural moisturizing mechanisms and it can lead to photosensitivity. Manufacturers use it because it is unbelievably cheap.

Propylene Glycol. Ideally this is vegetable glycerine mixed with grain alcohol, both of which are natural. Usually it is a synthetic petrochemical mix used as a humectant. It has been known to cause allergic and toxic reactions. (NOTE: Although not part of Aubrey's list, according to *The Cure for All Cancers*, also avoid anything with **propyl or iso-propyl alcohol** as it interferes somehow with the liver's ability to trap and kill parasites that play a role in cancer formation.)

Sodium Lauryl Sulfate. This synthetic substance is used in shampoos for its detergent and foam-building abilities. It causes eye irritations, skin rashes, hair loss, scalp scruf similar to dandruff and allergic reactions. It is frequently disguised in pseudo-natural cosmetics with the parenthetic explanation "comes from coconut."

Stearalkonium Chloride. A chemical used in hair conditioners and creams, it causes allergic reactions. Stearalkomium chloride was developed by the fabric industry as a fabric softener, and is a lot cheaper and easier to use in hair conditioning formulas than proteins or herbals, which do help hair health. Toxic.

Synthetic Colors. The synthetic colors used to make a cosmetic "pretty" should be avoided at all costs, along with hair dyes, as they are carcinogenic. They will be labeled FD&C or D&C, followed by a color and a number.

Synthetic Fragrances. The synthetic fragrances used in cosmetics can have as many as 200 ingredients. There is no way to know what the chemicals are, since the label will simply say "Fragrance". The potential problems caused by these chemicals include headaches, dizziness, rash, vomiting, skin irritation, etc.

Triethanolamine (TEA). It is used to adjust the pH and with many fatty acids as the base for a cleanser. TEA causes allergic reactions, including eye problems and dryness of hair and skin. It may be toxic with chronic exposure.

Disposal of Hazardous Wastes

A hazardous waste is defined as any substance that is toxic, flammable, corrosive, reactive or explosive when mixed with other substances. Many of the products you have in your kitchens, basements or garages right now probably fit one of these definitions. Americans generate 1.6 million tons of household hazardous waste per year. Unfortunately, most of this waste ends up contaminating our water, air and soil because people do not know how to dispose of these wastes properly. We pour them down the drain, on the ground, in a storm sewer, or throw them in the trash. Wastewater treatment plants and sanitary landfills are not equipped to handle most types of hazardous wastes and these wastes end up injuring sanitary workers and other people or wildlife.

Don't ever dispose of hazardous wastes by pouring them down the drain. If you have a septic system, drains lead to a series of buried tanks where liquid overflow passes into a set of perforated pipes filtering into the soil. Bacteria degrade most of the wastes in a septic system. Hazardous wastes may not only kill off the organisms that degrade the wastes, but also may contaminate your backyard soil or well system. Wastewater treatment plants are also not equipped to handle toxic wastes, and toxic wastes poured down the drain eventually end up in our environment.

The best way to deal with hazardous wastes is to substitute safer materials whenever possible. (See *Recipes* and *Safe Product Sources*.) If there is no substitute, buy only what you need and dispose of the container or any surplus responsibly. This may entail taking the leftover material, or even empty container, to the nearest recycling center, community collection point or licensed hazardous waste contractor. Contact your area's solid and hazardous waste department or federal environmental agency for information on the nearest appropriate location. (See *Further Resources*.) You might also consider donating leftovers to a business, charity or government agency. Theater groups often need paint, and a local green house or agricultural center may need pesticides, for example. The attached chart should help you determine the best means of disposal for typical, hazardous household products. (Unless the product is one of the safe alternatives, you can assume it contains hazardous materials, particularly if there is any kind of warning on the label. See *Labeling and Regulation*.)

Hazardous Waste Disposal Chart

Think before you throw it out! The following chart helps you determine where to dispose of various types of common household wastes. Except for some pesticides and medicines, the best way to dispose of any leftovers is to give them to someone who can use them up. The second best options are described by the following key: X - Give the waste to a hazardous-waste collection program. R - Recycle the waste. D - Flush down the drain with lots of water. (Rinse bottles.) T - Place in the trash.

Type of Waste	Disposal	Type of Waste	Disposal
Biocides		*Cleaners/Housekeeping*	
Ant and Roach Killer	X	Ammonia-Based Cleaners	D
Flea Powder/Dip	X	Abrasive Cleaners	D
Fungicide	X	Chlorine Bleach	D
Insecticide	X	Drain Cleaners (with lye)	D
Nematicide	X	Disinfectants	D
Molluscicide	X	Floor and Furniture Polish	X
Rat Poison	X	Metal Polish (with solvents)	X
Weed Killer	X	Moth Balls	X
		Oven Cleaners (with lye)	D
Painting and Woodworking		Rug/Upholstery Cleaner	X
		Shoe Polish (dried)	T
Brush Cleaner (with TSP)	D	Shoe Dye	X
Brush Cleaner (with solvents)	X	Spot Removers	X
Glue (solvent-based)	X	Toilet Cleaners (with lye)	D
Glue (water-based, dried)	T		
Furniture Stripper (with lye)	D	*Hobbies*	
Furn. Stripper (methylene chloride)	X		
Latex/Water-Based Paint (dried)	T	Artists' oils/acrylics - most (dried)	X
Oil-Based Paint	X	(yellows, reds, green, white, violets)	X
Thinners and Turpentine	X	Batteries	X
Wood Preservatives or Stains	X	Chemistry Set	X
		Ceramic Glaze (solidify by firing)	T
Automotive Supplies		Glue (solvent-based)	X
		Gun Cleaning Solvent	X
Antifreeze (ethylene glycol)	X	Model Paint	X
Antifreeze (propylene glycol)	D	Pool Chemicals	X
Auto Battery	R	Photographic Chemicals	X
Brake Fluid	R or X		
Car Wax/Chrome Polish	X	*Bathroom*	
Engine Degreaser	X		
Gasoline or Diesel Fuel	X	Cosmetics and perfumes	D
Transmission Fluid	R	Medicines	D
Used Engine Oil	R	Nail Polish (dried)	T
Windshield Washer Solution	D	Nail Polish Remover	D

Labeling and Regulation

Consumers should be aware that they cannot rely on labeling to inform them completely about potential ingredients and hazards associated with the use of any particular product. Even with all the rules and regulations, many problems are associated with product labeling. The following illustrates some of these:

o *Warnings, such as "danger" and "poison," are required only to inform consumers of immediate or acute dangers associated with the use of a product, and not its long-term or chronic health effects.* Most often there is no indication of potential cancer-causing ingredients or chemicals that may contribute to birth defects, kidney or liver damage or nervous system damage, etc. Many people also experience chemical sensitivities after exposure to certain glues, paints, pesticides or other strong chemicals. (A chemical sensitivity means a collapse of the immune system so that any exposure to most commercial products produces a strong reaction.) Though not yet chemically sensitive, many people are not aware that their allergies, rashes, headaches, asthma or other symptoms may be caused by the products they use daily.

o *Labels are not required to list all product ingredients.* Manufacturers are only required to list all active ingredients, and ingredients that are above a certain concentration. Sometimes, some of the "inert" substances found in products are more hazardous than the active ingredients when you consider long-term chronic impacts such as carcinogenicity, neurotoxicity, mutagenicity or teratogenicity. Even if a product contains a chemical that may cause an acute or immediate problem, it may not be listed if it is below a certain concentration. Methyl alcohol, for instance, can cause blindness if ingested, but does not require labeling if it is being used in quantities of less than four percent.

o *A label stating that a product is nontoxic or natural does not necessarily mean that it is absolutely safe or found in nature.* Federal regulation defines substances as nontoxic if less than fifty percent of lab animals die within two weeks while being exposed to the ingredient or product through ingestion or inhalation. This means that a product can still be labeled nontoxic if up to fifty percent of lab animals die within two weeks. Long-term and chronic effects are not considered. No legal definition of natural is available yet, so a product is frequently labeled natural if it contains some natural ingredients. It may still contain many fillers or substances made from petrochemicals which may not be healthy on a long-term basis. (See *List of Cosmetic Ingredients to Avoid*, for example.)

o *Product labels do not always inform consumers of the type of hazard associated with the product or give accurate first aid recommendations.* Products do not always have an appropriate warning for those at high risk. Methylene chloride, for example, interferes with the blood's ability to carry oxygen to body tissues. Yet products containing this compound – paint strippers, for example – do not carry special warnings for those with heart or blood conditions. Moreover, a random survey conducted by the New York Poison Control Center found eighty-five percent of labels from over 1,100 products had inadequate or erroneous first aid information. In emergencies, call your closest Poison Control Center.

Natural Lawn and Garden Care

Natural Fertilizers. The best fertilizers are those that give the full range of minerals in the same proportions as nature. These would be harvested from natural sources, rather than synthesized, like minerals harvested from ancient seabed, clay mineral deposits, or the rock-dust ground from local mixed gravel or granite. Other natural fertilizers include animal manure (chicken, turkey), bat guano (should be certified to be free of rabies), bone meal, fish emulsion and meal, kelp meal, lime and oyster shell, hoof and horn meal, bracken ferns, wood ash (also good for pest control) and soft phosphate. Compost and manures are natural soil amendments.

Compost. Compost improves your soil's texture and structure, helps retain essential nutrients, control erosion and regulate soil moisture, regulate pH and protect against weeds. You can buy compost, but making it at home is best and provides a way to recycle your organic matter instead of adding it to already overloaded landfills. It is easier than you think to compost. Novices should start with fallen leaves and cut grass, as composting food waste is where most beginners have some problems. Leaves are a concentrated source of nutrients, carrying fifty to eighty percent of the nutrients that a tree extracts from the soil and air; and grass is a rich source of nitrogen. First, make a base to lift the pile off the ground so that air can circulate from the bottom to the top. This way you only have to turn the pile three to four times a year. Wear a mask when turning the pile to protect yourself from mold spores. To make a base, get a wooden pallet (usually free from a hardware store or post office) or make it with piled up sticks. The bin itself can be made of wire caging cut in 12 1/2 foot lengths, wrapped around into a circle and tied with wire ties. A fifty foot length will make four and costs about $14 to $25. Fill your cage with organic materials. When filling it with leaves, be sure to wet well between each layer to pack them down. A few dos and don'ts make it easier. Don't try to compost grass by itself. Don't put in sticks larger than your thumb (the rule of thumb). Never try to compost meat or dairy products (except crushed eggshells), or anything that has been processed with oils. Always bury food under one foot of material and chop it up first. Your pile should never stink. If it does, something is wrong. Don't compost dandelions or other weeds that have gone to seed (unless you want a yard full). Other items that don't compost well include: magnolia leaves, corn cobs, thorns, holly, English ivy, pine cones and sweet gum balls. Don't compost pine straw as it is already nature's perfect mulch. You can compost cardboard anything as long as you soak it in water a couple of days first. You don't have to worry about adding worms; build it and they will come.

Natural Pest Control. If, after restoring the health of the soil, you still have pest problems, there are several natural ways to handle unwanted animals, insects and plants. These include traps, companion plants that deter harmful insects, beneficial insects and organisms and natural minerals, soaps, oils and botanical pesticides. Natural predators include praying mantises, beneficial nematodes (*Neoplectana* and *Heterorhabditis)*, milky spore (for Japanese beetles), lacewing larvae, lacewings and ladybugs and predator snails (ask your local nursery). Fatty acid soap sprays help with many kinds of pest infestations. Sulfur dusts help with fungi, and diatomaceous earth helps with snails. Pyrethrins and other natural botanicals also work well for garden pests, but should be used with care and as a last resort.

General Environmental Issues

We live in an unprecedented time in human history. Environmental challenges have never been greater, but the opportunities to meet these challenges are equally great. In this short section, I describe some of the important challenges we currently face. If many scientists, sages and indigeous peoples are correct that we are all connected, not only with each other, but also with all living creatures and the earth itself, global issues are the concern of each and every individual.

The growth in the human population is our most accute problem today. Most of our other problems, at the current level of consciousness and technology, stem from this. Our numbers, only 500 million at the time of Columbus and 1.5 billion at the turn of the century, are rapidly reaching six billion. They may reach ten billion or more in the next forty years, well within the lifetime of most of us.

The way we have chosen to live also creates difficulties. Our current agricultural and industrial practices have begun to unravel the web of geochemical and biological life. I will only mention a few examples. (1) Due to the production of man-made chemicals, atmospheric chlorine levels have risen 600 percent in the last forty years. As a result of these and other chemicals, the ozone layer that protects the earth from too much ultraviolet radiation has been damaged, not just at the polar ice caps, but also right overhead. (2) The burning of gasoline and other fossil fuels, and the cutting of forests, has doubled atmospheric carbon dioxide (CO_2) levels – from 300 to 600 ppm – in just fifty years. Carbon dioxide acts like an atmospheric greenhouse, trapping heat. With just the $1°$ average rise to date, scientists have already noted some melting of the polar ice caps. Even mountain peaks traditionally continually covered with snow are beginning to melt down. The glacial Alps have melted enough to expose, for the first time in human history, a man who had been captured in the ice 5,000 years ago. Continued melting could cause a sea-level rise, changing the face of the globe as we know it today.

(3) In the biological realm, most of the Earth's ecosystems were fairly intact at the time of Columbus. Today, nearly forty percent of the Earth's surface has been converted to agriculture, pasture, city or other human use. This may not sound like much, but the important factor has been the rate of change. The vast majority of the change has occurred in the last fifty years with the exponential growth of the human population. For example, in Central America, more than two-thirds of the post-Columbian deforestation occurred after 1950. Even in 1970, still one half of the forest remained. Now, the region has less than one-third forest cover remaining and deforestation continues. At last count, this region and similar regions in the world were losing 10.4 million hectares of tropical forest a year, an area about the size of the state of Colorado.

(4) As most of the world's good agricultural land has already been taken, this spread of human populations is taking place on hillsides where soil erosion is destroying the land, rivers, estuaries and coral reefs, not to mention roads, irrigation canals and the capacity of dams. (5) People are also spreading into areas with an unusual diversity of plant and animal life, like the tropical rain forest. Even ten years ago, biologists estimated that some sixty-eight percent of original wildlife

the Far East had disappeared, and some sixty-five percent in Africa. The earth is losing plant and animal species at an unprecedented rate. The few extinction events that occurred in geological history took tens to hundreds of thousands of years to occur. This time, in one human lifetime, we may destroy as many as a fifth to a fourth of all lifeforms on the planet. Although the earth has seen vast ecological changes in its history, never before has it been through geochemical, physical and biological changes on such a vast scale in such a geologically short period of time.

Although the challenges seem nearly overwhelming, I personally believe that with God's help they can be resolved. We seem to be on the brink of catastrophe, but we are also on the brink of incredible technological and psycho/social change. Many of these changes seem to stem from our discoveries in quantum physics and the radical change in world view that these discoveries engender. In particular, the proposed existence of a Unified Field at the basis of creation has reshaped the global perspective and brought Western science and Eastern spirituality to common ground. The marriage of understanding is producing a deluge of creativity and human advancement.

Technological Signs. Although still largely unrecognized, technological breakthroughs have occurred that will radically affect our environment and the way we live. For instance, a machine that reaches temperatures of $10,000°$ F + and can break down any toxic waste to its elemental particles now exists. More amazing is that this invention puts nothing in the atmosphere. It does not burn wastes, but thermally breaks them down. It has an extremely high rate of destruction and amazingly low operating costs. Other exciting breakthroughs include: (1) a device that makes it possible to generate all the electricity that you need in your home on-site and pollution-free; (2) an invention to retrofit your car to operate on water; (3) a device to allow operation of all lighting and appliances off DC current right in a business or home. Finally, (4) the UT code is a discovery based on Unified Field Theory that allows us to eliminate electro-magnetic "noise" or incoherence and the resulting discomfort and symptoms associated with exposure to electromagnetic fields. (See *Electromagnetic Fields and Coherence*.) This discovery also increases the efficiency of electricity use. (5) This section would not be complete without mention of organic agriculture. Although still in the minority at the present time, organic farming that works cooperatively with nature is beginning to reach a certain maturity and spread throughout the world. Signs indicate that even the American lawn will be affected. New breakthroughs in grass seed that flourishes under totally organic lawn care techniques now exist. (See *Product Sources*.)

Psycho-Social Signs. Human beings seem to be awakening to the possibilities of their own potential in great numbers. One only has to turn on the television set and witness the unprecedented programming on miracles and other unexplained phenomena. Books and speakers on how to achieve higher states of consciousness and its implications have proliferated. Some scientists theorize a link between the unified field described by theoretical physics and the field of human consciousness. Teachers show us how to tap into the "super-conscious" to resolve human problems. A kernel of individuals in business are practicing new modes of cooperation and mutual assistance, rather than mistrust and competition. These human models are still small but spreading. (See *Further Resources*.)

Cosmic Consciousness

This wonderful description of C.C. is by Denise Linn from her book *Sacred Space*. Her experience occurred when, as a teenager, she was shot by a stranger.

"At the hospital, everything seemed amplified. The lights appeared glaring and bright. Searing pain. Shrill harsh voices. Slowly, the lights began to dim. Pain subsided. Voices faded into stillness. I found myself in a soft womblike darkness. I felt as if I were being drawn deep down within a velvet black cocoon.

Instantly the black bubble seemed to burst. The most brilliant luminous golden light enveloped me. It was so vibrant that the brightest sun would pale in comparison. Everywhere around me, into infinity was light. Infused in the light with crystalline delicacy was pure sweet music. This liquid light symphony was ebbing and flowing throughout the universe in perfect harmony. The fluid harmonics pervaded my being until I merged with the light and sound. Light and sound were not separate from each other. I was light-sound. And this surrounding, all pervasive universe of warmth and light and music seemed completely natural *and completely familiar.*

Everything seemed more real than anything else I had ever experienced! It was as if my teenage life up to that time had only been a dream. Just as when you awake in the morning and your dream begins to fade in the "reality" of the day, my entire life up to that time seemed to dissolve into a fine mist as I stepped into this new "hyper-real" reality. My previous life seemed nothing more than an illusion to me.

All time seemed to flow in a continuous, everlasting "Now." There was no past and no future. Everything was contained within an infinite present. I remember trying to think of the past and I couldn't, because it was inconceivable. It literally didn't exist. It was as impossible for me to imagine linear reality when I was "there" as it is for me to fully understand non-linear reality when I am "here".

Completely infused within this world of light/sound/infinite-now was "Love." This was so very different from the way we usually think of love. Our culture's conception of love involves loving someone or something as an entity separate from ourselves. The love I experienced was infinite and limitless. There wasn't anything that wasn't Love. The love I experienced was not separate from anyone or anything. It was as natural as breathing. Everything simply *was* Love, a part of it, without any separation. It was love beyond form, without boundaries.

And I *wasn't alone.* You were there with me. Everyone was there. There wasn't anyone who wasn't there. We were all One. We weren't separate. There was no beginning, no end, just infinite, eternal light. No longer confined to my body, I experienced being one with all things and all beings. I was everyone that I had ever loved and everyone that I had ever hurt. I was everyone that I had ever known and I was everyone that I would never know. I was the hungry beggar on a side-street in Delhi. I was the thief in New York City. I was the baby held in her mother's arms in Kenya. I was the spiritual adept in a mountain temple in Japan. I was everyone and everyone was me."

From *Autobiography of a Yogi* by Paramahansa Yogananda. In this event, his guru strikes Yogananda gently on his chest above the heart and bestows upon him the experience of cosmic consciousness that he had passionately sought for so long.

"My body became immovably rooted; breath was drawn out of my lungs as if by some huge magnet. Soul and mind instantly lost their physical bondage and streamed out like a fluid piercing light from my every pore. The flesh was as though dead; yet in my intense awareness I knew that never before had I been fully alive. My sense of identity was no longer narrowly confined to a body but embraced the circumambient atoms. People on distant streets seemed to be moving gently over my own remote periphery. The roots of plants and trees appeared through a dim transparency of the soil; I discerned the inward flow of their sap.

The whole vicinity lay bare before me. My ordinary frontal vision was now changed to a vast spherical sight, simultaneously all-perceptive. Through the back of my head I saw men strolling far down Rai Ghat Lane, and noticed also a white cow that was leisurely approaching. When she reached the open ashram gate, I observed her as though with my two physical eyes. After she had passed behind the brick wall of the courtyard, I saw her clearly still.

All objects within my panoramic gaze trembled and vibrated like quick motion pictures. My body, Master's, the pillared courtyard, the furniture and floor, the trees and sunshine, occasionally became violently agitated, until all melted into a luminescent sea; even as sugar crystals, thrown into a glass of water, dissolve after being shaken. The unifying light alternated with materializations of form, the metamorphoses revealing the law of cause and effect in creation.

An oceanic joy broke upon calm endless shores of my soul. The Spirit of God, I realized, is exhaustless Bliss; His body is countless tissues of light. A swelling glory within me began to envelop towns, continents, the earth, solar and stellar systems, tenuous nebulae, and floating universes. The entire cosmos, gently luminous, like a city seen afar at night, glimmered within the infinitude of my being. The dazzling light beyond the sharply etched global outlines faded slightly at the farthest edges; there I saw mellow radiance, ever undiminshed. It was indescribably subtle; the planetary pictures were formed of a grosser light.

The divine dispersion of rays poured from an Eternal Source, blazing into galaxies, transfigured with ineffable auras. Again and again I saw the creative beams condense into constellations, then resolve into sheets of transparent flame. By rhythmic reversion, sextillion worlds passed into diaphanous luster, then fire became firmament.

I cognized the center of the empyrean as a point of intuitive perception in my heart. Irradiating splendor issues from my nucleus to every part of the universal structure. Blissful *amrita*, nectar of immortality, pulsated through me with a quicksilverlike fluidity. The creative voice of God I heard resounding as *Aum*, the vibration of the Cosmic Motor."

Inner Housecleaning

A book about toxins in our environment is incomplete without mention of what is probably the most significant and widespread source of toxicity in our lives – our own unproductive thoughts and emotions. Toxic thoughts and emotions are not just a figure of speech. A growing body of research reinforces the mind/body medical paradigm and shows that thoughts and the resulting emotions generate measurable physiological changes which can powerfully affect either healing or the development of disease. Whether emotions are directed at ourselves or others, the negative or positive results in the body are the same. The body does not distinguish!

We have far more control over our thoughts, our resulting emotions and our actions, and therefore our lives, than we may realize. Most of our problems stem not from external events, but from our interpretations of them. These interpretations are produced by our belief systems, both conscious and unconscious, which grew out of our life (particularly childhood) experiences. Mature and well-adjusted people have largely liberated themselves from the trap of their past and any non-functional beliefs. They are not dependent upon outside influences for their happiness, but instead have great wellsprings of creativity and inner resourcefullness, including faith in a higher power, to respond to any situation. Maharishi calls this process "self-referral." Different modalities can help us to become more inner-directed. These include meditation, prayer, therapy, hypnosis, Edu-K (to remove emotional blocks), visualization, simply acting "as if" or in spite of fear, or even noticing the thoughts that precede strong emotions, writing them down, and then writing down more rational responses. I say that we are happy to the degree that we take responsibility for and feel "response ability" to our current situation. We are also happiest in pursuit of our own goals, as part of a larger purpose, knowing our power and creative options. The following are references that I found life-changing. (See also *Further Resources.*)

Allen, James, n/d. *As a Man Thinketh.* Marina del Rey, CA: DeVorss & Co. and Hulst, Dorothy J., n/d. *As A Woman Thinketh.* Same publisher.

Burns, David, 1980. *Feeling Good, The New Mood Therapy.* New York: Penguin.

Cutright, Layne and Paul, 1996. *Straight from the Heart.* La Costa, CA: Heart to Heart.

Gray, John, 1992. *Men Are from Mars, Women Are from Venus.* New York: HarperCollins Publishers, Inc.

Hay, Louise, 1987. *You Can Heal Your Life.* Carson, CA: Hay House, Inc.

Levine, Stephen, 1982. *Who Dies?* New York: Anchor Books, Doubleday.

Minto, Wally, 1994. *Communication & Understanding in Relationships.* Dallas, TX: THE ADvisor group, inc.

Peck, Scott, 1980. *The Road Less Traveled.* New York: Touchstone.

References

Adams, R.M. and H.I. Mailbach, 1985. "A five-year study of cosmetic reactions." *Journal of the American Academy of Dermatology* 13 (6): 1062-69.

Baker, B. and P. McGee, 1991. "CDFA pesticide residue analysis of organics." *California Certified Organic Farmers State Newsletter* 8 (4): 6.

Berthold-Bond, A., 1990. *Clean and Green. The Complete Guide to Nontoxic and Environmentally Safe Housekeeping.* Woodstock, NY: Ceres Press, 162 p.

Clark, H. R. PhD, ND, 1993. *The Cure for all Cancers.* San Diego, CA: ProMotion Publishing, 511 p.

Cantor, K.P. et. al., 1988. "Hair dye use and risk of leukemia and lymphoma." *American Journal of Public Health* 78 (5): 570-71.

Carper, J., 1993. *Food-Your Miracle Medicine.* New York: HarperCollins Publishers, Inc., 528 p.

Committee on Interstate and Foreign Commerce, Subcommittee on Oversight and Investigations, 1978. *Cancer Causing Chemicals in Food. Ninety-Fifth Congress.* Washington, D.C.: GPO.

Cone, Virginia et. al., 1986. *National Body-Burden Database. Chemicals Identified in Human Biological Media. 1984.* Volume 7. Part II. Washington, D.C.: U.S. EPA, DOE and NIH. EPA-560/5-84-003.

Dadd, D.L., 1990. *Nontoxic, Natural & Earthwise. How to Protect Yourself and Your Family from Harmful Products and Live in Harmony with the Earth.* New York, NY: Jeremy P. Tarcher/Perigee Books, 360 p.

Davis, D. et. al., 1994. "Decreasing cardiovascular disease and increasing cancer among whites in the U.S. from 1973 through 1987." *Journal of the National Cancer Institute* 271 (February): 431-37.

Davis, J.R. et. al., 1993. "Family pesticide use and childhood brain cancer." *Archives of Environmental Contamination and Toxicology* 24: 87-92.

Dix, Karen, 1990. *Cleaning Brochure. Safe and Natural Recipes to Clean Your Home or Office!* Conowingo, MD: Karen's Nontoxic Products. Also 1994 course.

Federal Registers 44 (No. 70), 47 (Nos. 108, 188, 224), 50, 51 and 53 (No. 168).

Foster, H.D., 1992. "Aluminum and health." *Journal of Orthomolecular Medicine* 7 (4): 206-8.

General Accounting Office (GAO), 1991. *Food Safety and Quality.* Washington, D.C.: GAO.

Golden Empire Health Planning Center, 1990. *Making the Switch. Alternatives to Using Toxic Chemicals in the Home.* Sacramento, CA: The Local Government Commission, 41 p.

Graves, A.B. et. al., 1990. "The association between aluminum-containing products and Alzheimer's disease." *Journal of Clinical Epidemiology* 43 (1): 35-44.

Greenfield, E. J., 1987. *House Dangerous. Indoor Pollution in Your Home and Office – and What You Can Do About It!* New York, NY: Vintage Books, 225 p.

Hampton, A., 1994. *Natural Ingredients Dictionary. Plus 10 Synthetic Cosmetic Ingredients to Avoid.* Tampa, FL: Organica Press, 31 p.

Hardell, L. et. al., 1981. "Malignant lymphoma and exposure to chemicals, especially organic solvents, chlorophenols, and phenoxy acids: a case-controlled study." *British Journal of Cancer* 43: 169.

Harlow, B.L., 1992. "Perineal exposure to talc and ovarian cancer risk." *Obstetrics and Gynecology* 80 (1): 19-26.

Harrison, N., 1988. "Migration of plasticizers from cling-film." *Food Additives and Contaminants* 5 (1): 493-99.

Hoar, S.K. et. al., 1986. "Agricultural herbicide use and risk of lymphoma and soft-tissue sarcoma." *Journal of the American Medical Association* 256: 1141.

Hunter, L.M., 1989. *The Healthy Home: An Attic to Basement Guide to Toxin-Free Living.* New York, NY: Pocket Books, Simon & Schuster Inc., 313 p.

International Agency for Research on Cancer (IARC), 1987-91. *IARC Monographs on the Evaluation of Carcinogenic Risks to Humans.* Vols. 1-51. Lyon, France: World Health Organization (WHO).

Longnecker, M.P. et. al., 1988. "A meta-analysis of alcohol consumption in relation to risk of breast cancer." *Journal of the American Medical Association* 260 (5): 652-56.

Lowengart, R.A. et. al., 1987. "Childhood leukemia and parents' occupational and home exposures." *Journal of the National Cancer Institute* 79 (1): 39-46.

Morris, R.D. et. al., 1992. "Chlorination, chlorination by-products and cancer: a meta-analysis." *American Journal of Public Health* 82 (7): 955-63.

National Research Council, 1993. *Pesticides in the Diets of Infants and Children.* Washington, D.C.: National Academy Press, 386 p.

N/A, 1996. "Another meaty link to cancer: Red meat consumption may increase the risk of non-Hodgkin's Lymphoma." Science News 149 (23): 365.

National Toxicology Program, 1990. *Toxicology and Carcinogenesis Studies of Sodium Fluoride in F344/N Rats and B6C3F1 Mice (Drinking Water Studies)*. Research Triangle Park, NC: NIH Publication no. 91-2848.

Peters, J.M. et. al., 1994. "Processed meats and risk of childhood leukemia." *Cancer Causes and Control* 5: 195-202.

Savitz, D.A. et. al., 1990. "Magnetic field exposure from electric appliances and childhood cancer." *American Journal of Epidemiology* 131 (5): 763-73.

Sax, N. Irving and Richard J. Lewis, Sr., 1987. *Hazardous Chemicals Desk Reference*. New York: Van Nostrand Reinhold, 582 p.

Schuphan, W., 1974. "Nutritional value of crops as influenced by organic and inorganic fertilizer treatments—results of 12 years of experiments with vegetables." *Qual. Plant, Plant Foods Hum. Nutr.* 23: 333.

Smith, B., 1993. "Organic foods vs. supermarket foods: elemental level." *Journal of Applied Nutrition* 45 (1).

Stanley, John S., 1986. *Broad Scan Analysis of the FY82 National Human Adipose Tissue Survey Specimens*. Washington, D.C.: EPA, EPA-560/5-86-035.

Steinman, D. & S.S. Epstein, MD, 1995. *The Safe Shopper's Bible. A Consumer's Guide to Nontoxic Household Products, Cosmetics, and Food*. New York, NY: MACMILLAN, A Simon & Schuster Macmillan Co., 445 p.

Sternglass, E.J. and J.M. Gould, 1993. "Breast cancer: evidence for a relation to fission products in the diet." *International Journal of Health Services* 23 (4): 783-804.

Thomas, John, 1995. *Young AGAIN! How to REVERSE The Aging Process*. Kelso, WA: Plexus Press, 384 p.

Tufts University Diet and Nutrition Letter: *A link between red meat and prostate cancer?* May 1994 Volume 12 (3): 7-8.

Walker, M., 1993. "Aluminum-contaminated drinking water, milk, tea and cookware." *Townsend Letter for Doctors* (April): 288-92.

Wallace, Lance A., 1987. *The Total Exposure Assessment Methodology (TEAM) Study. Project Summary*. Washington, D.C.: U.S. EPA, EPA-600/S6-87/002.

Weathersbee, P.S. et. al., 1977. "Caffeine and pregnancy: a retrospective study." *Postgraduate Medicine* 62: 64-69.

Zahm, S.H. et. al., 1992. "Use of hair coloring products and risk of lymphoma, multiple myeloma, and chronic lumphocytic leukemia." *American Journal of Public Health* 82 (7): 990-97.

About the Author

Maryla Webb Records, M.F.S., is an environmental specialist with a Masters of Forest Science (M.F.S.) in Ecology from the Yale School of Forestry and Environmental Studies and a B.S. in Chemistry from the University of Alabama (Phi Beta Kappa). She has also completed graduate work in organic chemistry and worked for two years in a cancer lab at the University of Alabama in Birmingham until she decided on an environmental career in 1980. The largest part of her work has been in the field of sustainable development. She has worked as an international environmental consultant in program and project development, review and assessment for clients such as the World Bank and the Inter-American Development Bank. Her concern for the rising use of potentially dangerous chemicals and its implications for individual and global health led her to write this book.

About GREEN GLOBE PRODUCTS, Inc.

Green Globe Products is a company that sells a comprehensive line of products for healthier living, principally through mail order and on the WEB. It was formed with the recognition that people would not take the time to make use of products for healthier living unless they were affordable, understandable and easily accessible. The need for a relatively simple but comprehensive collection of environmentally conscious products formed a major motivation for our establishment.

Green Globe Products, Inc. grew out of the research conducted by Maryla Records in the preparation for this book. As she researched cancer, degenerative disease and aging, she also came across products and techniques to address these issues. She was particularly concerned about products for the use of herself and her family, as many people in her family had died of cancer, even during the preparation of the book. It took a couple of years to sort through all the types of products, product designs and manufacturers, diets and herbs.

It was well worth the trouble! We've learned valuable information to help you protect or enhance your health – information that is passed onto you in the products selected. The durable products choosen are well-made, effective, solidly built, competitively priced and produced by reputable and well-established manufacturers. These include a range of water purifiers, a shower filter, air purifiers, an electrostatic furnace filter, products to moderate electromagnetic fields and full-spectrum lighting.

Green Globe also carries some excellent "green" consumable household products that are free from harmful chemicals. They are ecological or even totally organic. For example, Helena Meyer skin care and make-up products are hand crafted from natural ingredients. The freshness of the herbs are the key to their effectiveness! Prices are reasonable considering the quality of the products and may even save you money over popular grocery store or department store brands.

Green Globe also offers some of the best nutritional and herbal supplements, all from select companies that use natural ingredients. Herbal preparations, such as those of Alive EnergyTM and Maharishi Ayur-VedTM, require exacting steps prescribed by ancient Chinese and Indian traditions. Freshness and effectiveness are again key criteria for inclusion in the product line.

You can join Green Globe and get discounts on all these products. The membership fee of $30 gives you five to fifteen percent discounts, free samples of selected products, and monthly news bulletins and discount specials on Email. If you would like more information or a catalog, just fill out and mail the inserted form. Write us at Green Globe, P.O. Box 723, Silver Spring, MD 20918 or Email: grglobe@erols.com. Also see our WEB site at http:// www.erols.com/grglobe/.

http://www.erols.com/grglobe/

Promoting Enhanced Life In
Accord with Natures Laws

We hope that you found the information in *Domestic 007* valuable.
To learn about membership in Green Globe Products, Inc. or request catalogs, please fill out and mail this brief questionnaire to Green Globe Products, P.O. Box 723, Silver Spring, MD 20918. Please enclose $3.00 for shipping and handling for each household where you would like a catalog to be sent. For the same info free, print from our WEBsite at www.erols.com/grglobe/. I am most interested in the following:
( ) water purifiers, ( ) air purifiers, ( ) full-spectrum lighting,
( ) electromagnetic field products, ( ) child safety products, ( ) household cleaners, ( ) personal care products, ( ) vitamins and nutritional products,
( ) herbal tablets and preparations, ( ) everything.

Name: _____

Organization: _____

Address: _____

City/State/Zip: _____

Phone: _____ FAX: _____

Email: _____

I purchased this book from: _____

Occupation: _____ Age: _____

No. of Children In Household : _____ Ages: _____

I have a friend that would like information as well.

Name: _____

Organization: _____

Address: _____

City/State/Zip: _____

Phone: _____ FAX: _____

Email: _____

Occupation: _____ Age: _____

Green Globe Products, Inc.
P.O. Box 723
Silver Spring, MD 20918

Enjoy your life and be happy.
Being happy is of the utmost importance.
Success in anything is through happiness.

More support of nature comes from being
happy.

Under all circumstances be happy, even if
you have to force it a bit to change some
long standing habits.

Just think of any negativity that comes at
you as a raindrop falling into the ocean of
your bliss.

You may not always have an ocean of bliss,
but think that way anyway
and it will help it come.

Doubting is not blissful and does not
create happiness.

Be happy, healthy and let all that love flow
through your heart.

Maharishi Mahesh Yogi

The more knowledge we acquire, the more mystery we find. A human being is part of the whole, called by us the Universe, a part limited in time and space. He experiences himself, his thoughts and feelings as something separate from the rest – a kind of optical illusion of his consciousness. This delusion is a kind of prison for us, restricting us to our personal desires and to affection for a few persons nearest to us. Our task must be to free ourselves from this prison by widening our circle of compassion to embrace all living creatures and the whole of nature in its beauty. Nobody is able to achieve this completely, but the striving for such achievement is in itself a part of the liberation, and a foundation for inner security.

Albert Einstein

Trees and animals, humans and insects,
flowers and birds;
these are active images of the subtle
energies that flow from the stars
throughout the universe.
Meeting and combining with each other
and the elements of the earth,
they give rise to all living things.

The wise person understands this,
and understands that one's own energies
play a part in it.
Understanding these things,
the wise one respects the Earth as mother,
and the heavens as father,
and all living things as brothers and
sisters.

Those who want to know the truth of the
universe should practice . . .
reverence for all life.

This manifests as unconditional love
for oneself and all other beings.

Taoist Reflections

A Great Gift

Could someone you know benefit from reading . . .

Domestic 007: Eliminating Hidden Killers in Your Home

ORDER FORM

Please send me _____ copies of *Domestic 007* at $15.95 plus $3.00 each for shipping and handling. Five percent of the profits from the sale of this book will be used to fund various environmental and social causes in support of life in balance with nature and in accord with natural law.

Name: _____

Organization: _____

Address: _____

City/State/Zip: _____

Phone: _____ FAX: _____

Email: _____

Autograph book(s) to _____

My check or money order payable to Earth-Wise Publications in the amount of $_____ is enclosed.

Mail to: Earth-Wise Publications
P.O. Box 723
Silver Spring, MD 20918

For quantity orders please write, call or FAX us for discount information.

Phone: (301) 681-3492, FAX: (301) 681-9892

Place
Stamp
Here

Earth-Wise Publications
P.O. Box 723
Silver Spring, MD 20918